Chaos Under Control

The Art and Science of Complexity

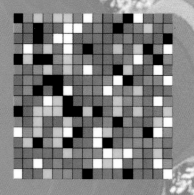

David Peak · Michael Frame

Chaos Under Control

The Art and Science of Complexity

Chaos Under Control

The Art and Science of Complexity

David Peak

Michael Frame

W. H. Freeman and Company
New York

Library of Congress Cataloging-in-Publication Data

Peak, David, 1941 –
    Chaos under control : the art and science of complexity / David Peak,
    Michael Frame.
        p.   cm.
    Includes bibliographical references and index.
    ISBN 0-7167-2429-4
    1. Chaotic behavior in systems.   2. Computational complexity.
    I. Frame, Michael.   II. Title.
    Q172.5.C45P43   1994
    003'.7 — dc20   94-3092
                                                                                CIP

PRINTED IN THE UNITED STATES OF AMERICA

1 2 3 4 5 6 7 8 9 0   VB   9 9 8 7 6 5 4

Contents

Foreword

A YEAR AGO, I gave the closing address at the International Congress of Mathematics Education ICME-7, which was held in Quebec City. My title was "Fractals, the Computer, and Mathematics Education," and I would like to begin by repeating a few of my remarks.

Clearly, for better or worse, I have ceased to be alone in an observation, a belief and a hope that keep being reinforced over the years. The observation is that fractals — together with chaos, easy graphics and the computer — enchant many young people, which — in turn — makes them excited about learning mathematics and physics. The belief is that this excitement can help make these subjects easier to teach to teenagers and to beginning college students. This is true even of those students who do not feel they will need mathematics and physics in their professions. Would it be extravagant to hope that, starting with this piece of mathematics called fractal geometry, we could help broaden the small band of those who see mathematics as essential to every educated citizen, and therefore as having its place among the liberal arts?

In this book, David Peak and Michael Frame have taken a step in realizing this hope. They experimented with the presentation of these ideas for six years in classes at Union College, the University of Richmond, and Yale University. I invited Michael Frame to Yale in the spring of 1993, and watched closely, as did an increasing number of my colleagues. The unanimous conclusion is that the course and the text Peak and Frame have produced are effective introductions for nonscience students. The approach they have taken deserves to become part of the curriculum at Yale and at other colleges and universities across the country. Adopting it is bound to increase by leaps and bounds the pitifully small number of liberally-educated persons friendly to mathematics and science.

Benoit B. Mandelbrot   February 1994

"IT'S AN EXCELLENT PROOF, BUT IT LACKS WARMTH AND FEELING."

Foreword

Confessions of a Math-Anxious Artist

THE HUMOR in this cartoon works because it reminds us that mathematics and emotion are considered to be opposites, just as art is placed opposite to science. In *CHAOS UNDER CONTROL: The Art and Science of Complexity*, David Peak and Michael Frame challenge these dichotomies and do for the reader something Mrs. Baumgartner, my sixth-grade math teacher, never did for me: they describe a mathematics that relates to our daily lives. How exciting and different this is from the tradition of mathematics as a series of problems to be solved (as if we didn't all have problems enough already!).

Until recently, I could never have imagined using mathematics as inspiration for my work as an artist. I can see now that I was for years in complete "math denial." It was *other* people — scientists and mathematicians, and other artists — who were literally painting cubes, squares, triangles, and other shapes, who used geometry in their work, not me. It didn't take much to unmask the fallacy in my reasoning; upon reflection, even the decision of whether to paint in two dimensions or to sculpt in three was a geometric one, not to mention the rules of composition and proportion we all follow as artists. I never understood that in this sense I was a geometric artist, using geometry in my work every day.

Similarly, you, the reader, may be "in denial," but I encourage you to let David Peak and Michael Frame give you a tour of the fascinating and irregular world in which we live. This is not a world of cold, abstract numbers; rather, it is familiar to us — fluctuations in the stock market or feedback noise from experiments with our camcorders and video machines — the irregular shapes and events that make up much of our daily lives. The ideals and averages that serve science well on a practical basis may seem sterile, a kind of unreal perfection incompatible with our human limits, but now we see that science has limits, too. That science can be overwhelmed by nature's "messiness" seems like an admission of vulnerability to us — and a bridge to our own.

Many popular and technical books have been written about complexity, chaos, and fractals. I have read most of these books — indeed, I collect them — because of my interest in the field; but, like the beds in *Goldilocks and The Three Bears*, some are too hard and some are too soft. Peak and Frame seem to get it *just right* for the general reader.

This new vision of the world as fractal and nonlinear is as inspirational and useful to me as it is to people from a staggering variety of fields and to people with varying levels of expertise. After finishing *Chaos Under Control*, the nonspecialist, in particular, will come away not as a passive reader but as a knowing participant, painlessly sharing the vision of the technical experts "in the know" at the frontiers of art and science.

Rhonda Roland Shearer   February 1994

Preface

The technical language of complex systems — fractals, chaos, the Butterfly Effect, self-organization, emergent properties — appears with increasing frequency in the mass media and in ordinary discourse. New products based on related ideas are gradually coming to the marketplace. Though the principles of complexity are not yet central to the syllabi of standard high school and college mathematics and science courses, their practical importance and intellectual appeal will surely soon result in their being taught side-by-side with trigonometric identities and Newton's laws of motion.

Over the past few years, numerous, generally well-informed and curious, but not technically trained, acquaintances have asked us, "Can you recommend a good, not too high-brow introduction to this stuff?" Our purpose in writing *Chaos Under Control* was to allow us to respond affirmatively to this question.

The shelves of any decent contemporary bookstore contain many books with references to chaos, fractals, and complexity in their titles. (A brief description of some of these, References 1 – 39, along with a few other introductory readings, References 40 – 58, is contained at the end of this book.) These include, on the one hand, some clearly written, quantitatively intensive texts. In most cases, the demands on the facility of the reader made by these books are well beyond what the vast majority of those asking us about "this stuff" can or care to bring to the task. Moreover, the scope of these books is frequently fairly narrow. We are convinced that the study of fractals, chaos, and complexity has the potential to connect many areas of intellectual pursuit, both within the sciences and between the sciences and the arts and humanities, but elucidating this broad, interdisciplinary perspective is not the focus of many of the quantitative books.

On the other hand, there are also a number of very fine qualitative and historical accounts of these subjects, and the more colloquial flavor found in these would seem better suited to the nonexpert reader. Regrettably, we have found that purely qualitative treatments of technical material mask a flaw. While they can acquaint one with new terms and provide a sketch of the principles, in the end, the result is often frustratingly fuzzy: an ability to use the new ideas with some precision and criticality requires a more engaged encounter with at least some of the details.

We have attempted, then, to write a book that stands somewhere in the middle ground. *Chaos Under Control* results from our experience, over the past five years, teaching the principles and applications of fractal geometry and chaotic dynamics to hundreds of beginning students at Union College, Yale University, and the University of Richmond. Our students typically had decent backgrounds in high school science and mathematics, but few, if any, were majoring in a technical discipline at the time they enrolled in the course, and many were self-avowed "math phobes." Thus, the material of this book has been subjected to extensive field testing by a large number of intelligent, though not quantitatively predisposed, readers. While emphasizing the qualitative and interconnecting aspects of the material, we have included in the book enough formal detail for the reader willing to grapple with a bit of algebra to emerge with a genuine functional mastery of the concepts (These detailed discussions are set off within the text by Mandelbrot Set icons like this: ———◆——— and ———◆ ◆——— , which indicate the beginning and the end of each discussion, respectively). To assist this effort, we have included some opportunities to explore further in a section called "Explorations and Challenges." These appear in various forms: mathematical puzzles, provocative questions, and hands-on experiments (done at least as well in a kitchen as a fancy lab).

Must you read *Chaos Under Control* in a linear fashion from beginning to end? No! That would run counter to the nature of the nonlinear phenomena we describe. Chapter One presents a quick introduction to most of the ideas in the book and probably should be read first. The focus of Chapters Two and Three is fractal geometry. Chapter Two, on iterated function systems, and Chapter Three, on fractal dimension, are independent of each other and are not required for any later chapter, except for the section "Iterated Function Systems" in Chapter Six. Chapters Four and Five deal with chaotic dynamics and are sequential. Chapter Six concerns the detection and control of chaos; it depends on both Four and Five. Except for the section "Recoding the Logistic Map," referring back to Chapter Five, Chapter Seven, on the Mandelbrot Set, is independent of the preceding chapters. Chapter Eight, which explores fractal basin boundaries, is better read after Chapter Seven, but is also reasonably independent. Chapter Nine, on self-organizing phenomena, stands pretty much alone, with the exception of one section, "Life, Sandpiles, and $1/f$ Noise," which requires some knowledge of the language of Chapter Six. Chapter Ten touches on a few novel applications of fractals and chaos in areas outside of the physical sciences and engineering to give a sense of their growing impact in diverse disciplines; Chapter Ten can be read either as dessert or in conjunction with Chapter One as an extended overview.

Our debts of gratitude are many. First, we thank the Alfred P. Sloan Foundation for grants from its New Liberal Arts and Special Leave Programs to support the early development of this work; we gratefully acknowledge related grants from the National Science Foundation (ILI-8952199), the Pew Charitable Trust, the Apple Corporation, and Union College's Internal Education Fund. The kind hospitality of the faculties and administrations of Yale University and the University of Richmond provided friendly and stimulating environments for testing this material with students at those institutions.

For several years we have enjoyed the generous support of Benoit Mandelbrot. His contributions include arranging for and partially funding the year one of us spent at Yale, widely advertising the concept of a course based on this material as an effective introduction to quantitative thinking for students outside science, and, indeed, bringing much of this material into current scientific and popular culture.

We thank James W. Corbett for his inspiration and continuing encouragement.

Richard Beals and Roger Howe made helpful comments on an earlier version of this manuscript and expedited the offering of a related course at Yale University.

Through their participation in summer workshops and some assistance in their areas of expertise, several members of the Union

College community have helped us refine this material. In particular, we thank Professors Thomas D'Andrea, William Fairchild, Neal Mazur, Daniel Robbins, Ruth Stevenson, Cherrice Traver, and Anton Warde. We especially thank Brian Macherone for his invaluable counsel in software matters and for the many hours he spent producing computer code.

All teachers learn from their students. We are happy to have had the comments and criticisms from the many who watched this material evolve over the last half decade. For help with some aspects of Chapter Ten, special thanks are due Sean Maclean, Catherine Sandler, and George Steele.

We thank Jerry Lyons and the W. H. Freeman Company for their vision and courage in taking this project on. Among the many at Freeman to whom we are indebted, we specifically thank Connie Day for constant vigilance to lapsed style and pulled punch and Kay Ueno for just about everything.

Finally, but most important, we thank Terry, who painstakingly read the many preliminary variations of this manuscript, and Jean for their perseverance, tolerance, and loving support.

David Peak
Michael Frame
February 1994

Extended problem sets separated by chapter and section and identified by type (analytic, numerical, computer-based, experimental, and essay), computer software useful for exploration, laboratory instructions, and course syllabi are available from the authors for the cost of mailing and handling. Please write to David Peak, Physics Department, Union College, Schenectady, NY 12308, or send for details by e-mail to PeakD @gar. union.edu.

Introduction:
A First Word

DO THE FOLLOWING SOUND FAMILIAR?

"Well! Look who got up on the wrong side of the bed this morning." — *a mother to her moody teenager*

"Right now, it looks like the weekend is going to be dry and pleasant, great for picnics and going to the beach. Of course, I don't really have that much confidence in forecasts so far down the line." — *a local TV meteorologist on the Monday weather forecast*

"Most of the blue chip stocks closed substantially lower today. It took me a bit by surprise, but in retrospect, I think it was probably just a technical correction." — *a stockbroker speaking to a client*

Life is often complicated, sometimes exceedingly so. Much of our everyday experience is unexpected, apparently whimsical, seemingly beyond our control. On the other hand, we also commonly take for granted the long-term, reliable functioning of refrigerators, computers, and communication satellites. How is it that some aspects of our experience are regular, predictable, tamable, while others appear to be the outcome of some cosmic game of chance? Is the universe a crazy patchwork of phenomena, some understandable, some beyond explanation?

Those aspects of our experience that we feel are most "under control" are typically linked to the ideas of science and the products of technology. You know, of course, that scientific writing often contains lots of mathematics. Despite suspicions to the contrary, this use of mathematics is not meant to prevent the innocent reader from discovering profound, though perhaps dangerous, truths; rather, it is meant to convey precision and promote clarity. It is a remarkable and mysterious fact that at least some pieces of the universe (the explainable ones) are best described in the language of mathematics.

Until very recently, scientists have been accustomed to describing the world in terms of what can be called "smooth" mathematics. Smooth mathematics is the mathematics of continuous and unjagged structures: unbroken lines, curves, surfaces, volumes. It includes major portions of arithmetic, algebra, geometry, and calculus. Its roots are as ancient as human history. Galileo, the first more or less modern scientist, expressed a deep belief that the geometry of Euclid is the language in which the secrets of the cosmos are written. Newton invented calculus, in part, to relate Euclidean geometry formally to the description of continuously evolving processes. The Euclidean–Galilean–Newtonian vision of the structure and dynamics of the universe has often proved extraordinarily useful. In large part, it has propelled the machinery of technology. Not coincidentally, this vision permeates much of contemporary Western thought.

Is the universe then partly Newtonian — smoothly continuous, a predictable clockwork mechanism — and partly messy stuff — social and psychological behaviors, esthetics, emotions, spirituality, free will, random happenstance, and all that? No. Such a dichotomy is too clean. Physical things that we feel should be intrinsically Newtonian are often quite unruly. Storms and earthquakes and tidal waves and all kinds of accidents elude prediction. And the shapes of nature aren't exactly smooth, either. Cézanne, in instructing his students to "treat nature in terms of the cylinder, the sphere, and the cone" [59], preached pure Euclidean–Galilean–Newtonian dogma. But although cylinders, spheres, and cones provide a kind of first-order approximation to reality, they also miss the essence of the natural. As Benoit Mandelbrot — mathematician, economist, and scientist of many persuasions — has put it in his seminal book *The Fractal Geometry of Nature* [60],

This is not a cloud.

This is not lightning.

This is not a mountain.

This is not a tree.

Clouds are not spheres, mountains are not cones, coastlines are not circles, and bark is not smooth, nor does lightning travel in straight lines.

The wide availability in the last quarter of this century of high-speed computers with large reserves of memory is rapidly transforming how we understand our surroundings. New notions and techniques are beginning to supplant some of the most venerated ideas of science and applied mathematics. Instead of approximating the inherently fractured character of nature with smooth forms, these new methods grapple with fracture head-on. Still very much an infant, this new science of complexity promises to describe the universe in much more accurate and appropriate terms, yielding, in consequence, deeper understanding and more reliable prediction. It also promises to much more closely ally the physical world with that of the mind, unifying what was previously thought dichotomous.

Much of the popular literature dealing with the topics of this book has an awestruck, vaguely mystical tone. Our intent is quite the opposite: we seek to demystify. As we created this book, one of our working titles for it was *Order, Chaos, Art, and Magic* (OCAM, for short). The *Magic* in that title is not an invocation of the supernatural or of forces forever beyond our understanding. Our *magic* concerns the craft of the illusionist: entertaining, counterintuitive, enchanting, but always completely natural and completely understandable. Similarly, by *art* we do not mean a nonscientific category of learning, but rather an exploration of the roots of esthetic response in physical law. And our *chaos* is certainly not confusion and disorder. Our *chaos* is supremely organized, dictated by rules, even capable of being brought under control. Foremost, our book is about *order*—regularities, simple explanations (that is *this* OCAM's razor*). This book is intended to offer the reader more than a nodding acquaintance with terms and ideas. It is meant to provide a functional introduction to some of the most fascinating and potentially useful concepts of this new science.

* Occam's Razor is the maxim "What can be explained on *fewer* principles is explained needlessly by *more.*" It was a central methodological principle in the work of fourteenth-century philosopher and Franciscan, William of Occam (Ockham).

Art and Magic

WE START BY discussing a little art and magic. This introductory chapter is a peek into an exhibition hall filled with wondrous, perhaps delightfully baffling, displays. We want to reveal all the secrets behind these seeming puzzles and, at the same time, discuss a number of big ideas, but we'll get around to all of that presently. First, we want to take a rapid tour through the galleries of fractals, chaos, and complexity to get your imagination revved up. Later, we will try to convince you that the weird and wonderful tricks you will soon see here have practical implications as well and are worth learning more about.

Squeals and Squiggles

How many times have you heard someone suddenly interrupted by a shrill burst from a nearby loud-speaker while talking to an audience with the aid of a microphone? It's so familiar you even know a name for it: feedback. You can perform a simple feedback experiment of your own. Get a small microphone and a small speaker. (Really small and cheap ones can be purchased at almost any electronics store.) Connect the microphone with wires to the input of your stereo amplifier and the speaker to the output. Point the speaker toward the microphone, as shown in Figure 1.1. Turn on the amplifier and gradually increase the volume control. Soon the speaker will start making a racket. (Can

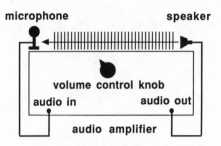

Figure 1.1   *Schematic set-up for audio feedback.*

you make the noise stop by turning the speaker away from the microphone?)

The squeal of *audio feedback* can be understood as follows: The speaker is designed to convert an electrical signal into vibrations in the air — that is, into sound. The microphone works in reverse. It converts vibrations in the air into electrical currents. To begin, the microphone "hears" some background sound from the room. In the first pass through the feedback loop, it sends an electrical signal to the amplifier, the function of which is to make that signal larger. The speaker, in turn, emits a sound that adds to the ambient room noise. The microphone picks up the new, louder sound and sends a stronger electrical signal to the amplifier than in the first pass. The amplifier consequently sends an even stronger electrical signal to the speaker, which, of course, responds by making even more noise. In short, through additional passes through the loop, the speaker begins to squeal violently. Such a repeated loop is said to define an *iterative process*. Iterative processes are a major source of complexity.

Here's another experiment you can probably do easily. Instead of using sound as the medium in which feedback is played, let's use light. In place of a microphone we'll use a video camera (a hand-held camcorder will do just fine), and for a speaker we'll use a video monitor; see Figure 1.2. (An interesting discussion of this experiment can be found in [61].)

When actually doing this experiment, you should probably first turn down the brightness control on the monitor to safeguard the camera from overexposure. Connect the camera output to the input of the monitor. (If your monitor is connected to

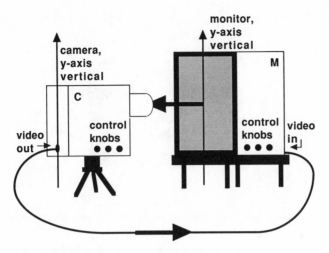

Figure 1.2 *Schematic set-up for video feedback.*

a VCR, connect the camera output to the input of the VCR.) Hold the camera upright (the y-axes of the camera and the monitor shown in Figure 1.3 are both vertical) about 4 feet from the screen and facing directly into it. Zoom until a slightly shrunken image of the whole screen appears. Once it does, you will see many such copies regressing ever more deeply into the "distance" (Figure 1.3). The center of the screen may be pretty bright. This brightness is roughly analogous to the amplification of sound in the previous experiment.

The video screen and our eyes permit the processing of much more information than do speakers and our ears, so *video feedback* can yield much richer results than audio feedback. To see some,

Figure 1.3 *The monitor screen and a few shrunken images within.*

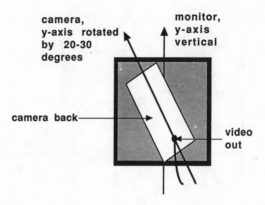

Figure 1.4   *View from behind the camera looking toward the monitor, after the camera is rotated.*

try rotating your camera by 20 or 30 degrees (see Figures 1.4 and 1.5). Turn off all lights in the room. If you see nothing on the screen, try turning up the brightness a bit. Eventually the screen will be illuminated by a flash. Depending on *all* the knob settings, any number of weird things can ensue. Least interestingly, you may obtain a stationary blob of light in the middle of the screen. Slightly more intriguingly, the blob may pulsate slowly, alternately growing and shrinking in size in a periodic fashion (Figure 1.6). If the knob adjustments are just right, all kinds of structured patterns may emerge. A few examples are shown in Figures 1.7–1.12 on pages 9–11.

The images shown in the figures are dynamic; they all show some variation in time. Some appear to be fairly periodic. Others never seem to repeat their appearances exactly. A remarkable

Figure 1.5   *The monitor screen and a few shrunken images within, after the camera is rotated.*

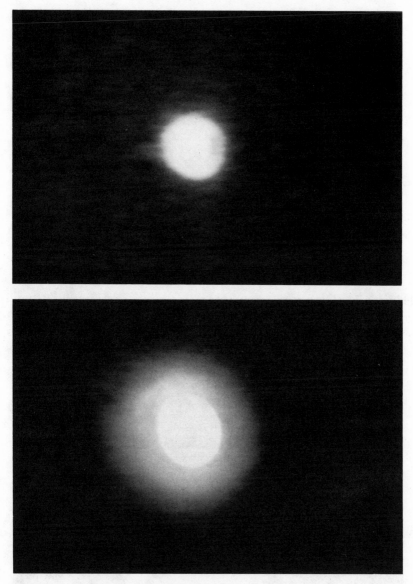

Figure 1.6   *A pulsating bright blob produced by video feedback. Two different instants are shown.*

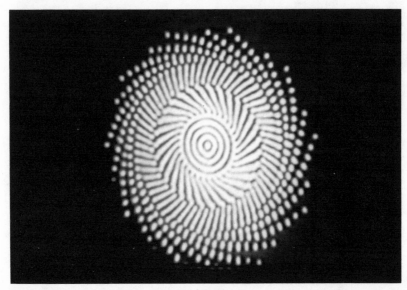

Figure 1.7 *This pattern generated by video feedback reveals varying degrees of ordered and disordered structure. For other patterns see Figures 1.8 – 1.12.*

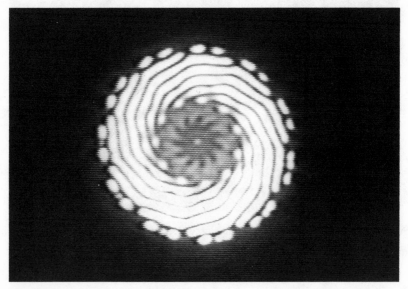

Figure 1.8

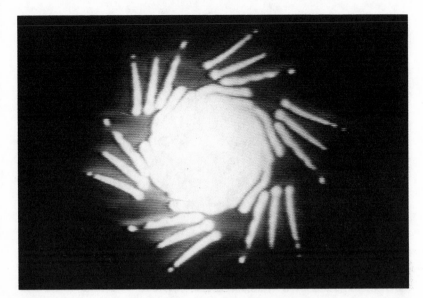

Figure 1.9

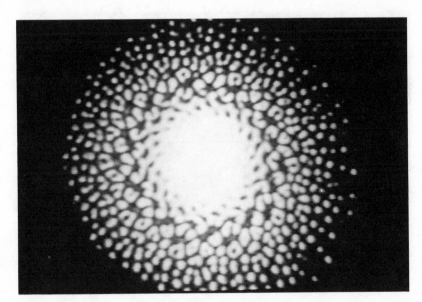

Figure 1.10

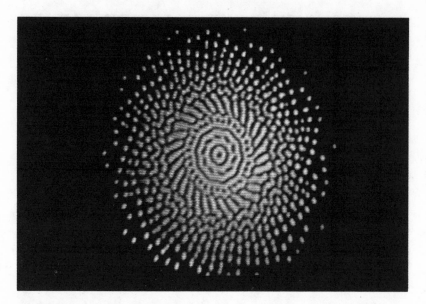

Figure 1.11

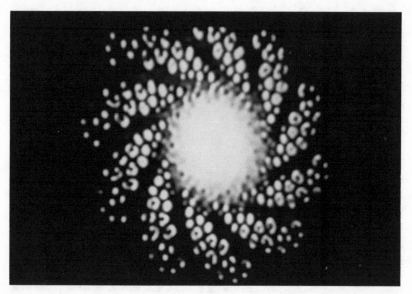

Figure 1.12

characteristic of many of the images is their robustness; if you try to alter the behavior by passing your hand between the monitor and the camera or by flashing the room lights on and off, very similar patterns often reappear after a short readjustment period. All of the images shown were produced by setting the color levels to produce a single-color (monochrome) picture. Multiple-color pictures are possible to obtain but are typically much less finely structured, more like whirling, blurry, colorful pinwheels (see Plate 1). The complexity and variation of video feedback images can be mesmerizing. If you don't immediately get satisfying patterns of light on your monitor, play around with the focus, color levels, and brightness and with the angle of rotation of your camera. Eventually you'll hit on some good combinations of settings.

A Pixel-Lighting Game

In a real video feedback set-up, light emitted by glowing phosphors on the monitor screen is focused by the camera's lens onto a sheet of semiconductor chips, which, in turn, generate flickering electronic currents. These currents, once delivered back to the monitor by connecting cables, create new sets of glowing phosphors, and so on. The details of the physics of the optoelectronics underlying what you see are pretty complicated, but the general features of what is observed in a real video feedback experiment can be reproduced without expensive lenses and electronic gizmos. The general features of interest are as follows:

1. If the brightness control on the monitor is set very low and light is introduced into the system via some external source (such as turning on the lights in the room briefly), the screen flashes, but the pattern produced rapidly extinguishes and thereafter the screen *remains dark.*

2. If the monitor brightness control is increased a little, the dim background glow of the unlit screen sometimes causes a spontaneous flash that rapidly stabilizes into an *unchanging or, perhaps, periodically pulsating blob.*

3. Sometimes, depending on all the knob settings, patterns do not stabilize at all. Instead, the screen flashes on, light dances across the

screen in some erratic fashion and disappears; a bit later, another flash leads to a different dance. This sequence of events *repeats only after very long intervals (if at all).*

4. Sometimes, again depending on focus, zoom, color levels, and so on, the spontaneous flash evolves into "organic" structures, *moving patterns with extraordinarily complex spatial structure.*

These features are typically robust with respect to perturbation. The nonquantitative impression one gets from playing with the knobs is that somehow the system is "underdriven" in behaviors 1 and 2 and "overdriven" in behavior 3. Behavior 4 is "just right," and finding the proper knob settings somewhere between underdriven and overdriven to produce "organic" patterns is a bit fussy.

We'd like to construct a kind of "thought experiment" to show that no fancy electronics are needed to produce the first two features. And we'd like to state, without support for now, that the third is just an extension of the first two. (Later on, we shall see that the fourth exhibits a form of self-organization, possibly modeling the origin of life!) Let's imagine a funny, puny, monochrome video screen consisting of a single row of 10 *pixels.* (A video monitor consists of a grid of tiny squares called "picture elements"—pixels, for short. A high-resolution monitor may have upwards of a million pixels. Each pixel on the screen can be dark or can be made to glow in different colors. Generally, the whole collection of pixels on a monitor is turned on and off— that is, the screen is "refreshed"—a few tens of times per second.) Each pixel on our thought screen is either on or off, and each is identified by a label, a number between 0 and 9.

Now, the game goes like this:

1. Draw a square.

2. Draw a diagonal from the lower lefthand corner to the upper right.

3. Draw a "single-hump" curve within the square so that one end is anchored at the lower lefthand corner, the other end is anchored at the lower righthand corner, and the hump is about halfway between.

4. Imagine that the diagonal is divided into the 10 pixels, pixel 0 at the lower left and pixel 9 at the upper right.

5. Start by putting a dot in the center of one of the pixels.

6. To find the next pixel, draw a vertical line from the starting point until it hits the curve, and from there draw a horizontal line until it hits the diagonal; that's the next pixel.

7. From the new point on the diagonal, draw a vertical line to the curve and a horizontal line to the diagonal to find the next pixel.

8. Repeat.

Figure 1.13 shows the schematic set-up. Drawn are four different situations corresponding to different "hump" heights. In the upper left version (A), one run of the pixel game is played to demonstrate what lines you have to draw. In the demonstration, the first pixel lit is pixel number 6; drawing a vertical line (down, in this case) to the curve and a horizontal line (to the left, in this case) leads to pixel number 2 lighting (and to pixel number 6 going off). Repeating leads to pixel 1, pixel 0 (or is it still pixel 1?—it's pretty close), pixel 0,. . . . If pixel 0 is interpreted as "all off," this sequence is like behavior 1 described in the list on

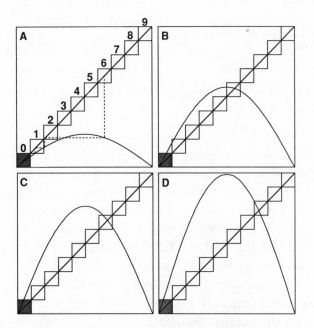

Figure 1.13 *Playing the pixel game with four different curves.*

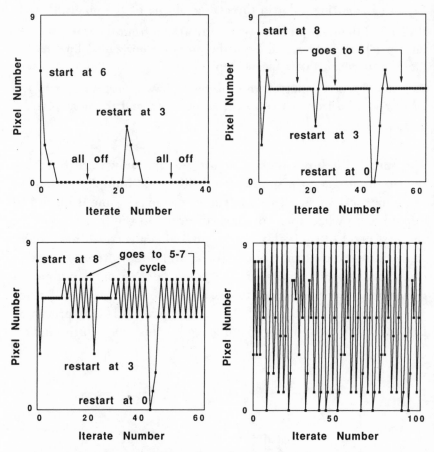

Figure 1.14   *Sequences of lit pixels in the pixel game for curves (a) A, (b) B, (c) C, and (d) D.*

pages 13 and 14. Try starting from another pixel; what you'll find is that they all lead to 0 for curve A.

Figure 1.14a shows a record of the sequences of lit pixels for curve A in Figure 1.13. In the first sequence, pixel 7 is the starting pixel; in the second, pixel 4. Both sequences terminate at pixel 0. You can think of the height of the hump in the game as being equivalent to the setting on a control knob in the video feedback experiment; different heights produce different behaviors in the game, just as different knob settings produce different behaviors on the screen. Figures 1.14b–d show what happens in examples B–D of Figure 1.13, respectively. (Try some by hand.)

In Figure 1.14b all the starting conditions lead eventually to a single lit pixel (5), while in Figure 1.14c all starting conditions lead to a periodic alternation between the same two pixels (5 and 7). This is type 2 behavior. Figure 1.14d, a sequence generated by example D, never repeats (type 3 behavior)! It's stunning, almost unbelievable, that such a complicated output can be produced by such a simple game. This behavior is called *deterministic chaos,* and we will have a great deal more to say about it later.

Another Pixel Game: Collages

The simple thought experiment in the previous section illustrates that no expensive equipment is necessary to make "pictures" that vary in interesting ways *in time.* All you need is the right kind of iterative scheme — one in which a new picture is made from an old one by the right mangling rule, and then a new new picture is made from what used to be the new picture by the same rule, and so on. Now, we show that interesting *spatial structures* can also emerge from simple iterative processing.

We can make pictures of only limited interest from the one-dimensional strip of pixels we just played with. More complex structures require two dimensions. The copies of the shrunken screens in Figures 1.3 and 1.5 suggest another thought experiment. Imagine a "video" set-up that iteratively generates new pictures from previous pictures as follows: Multiple copies of the previous picture are made, all shrunken and perhaps rotated and placed at different parts of the screen; the new picture results from "gluing" all these copies together. That is, it is a collage of the copies.

As a specific example, suppose the shrinking and placement rules are that one copy is the previous picture reduced by a factor of $\frac{1}{2}$ in each direction; that a second copy, identical to the first, is placed some distance horizontally to the right of the first; and that a third copy, also identical to the first, is placed some distance vertically above the first. The three copies are then collaged, yielding the new picture. Figure 1.15 demonstrates how this works when the starting picture is a square 100 pixels

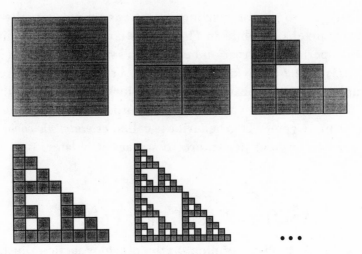

Figure 1.15   *Iterative processing of a square.*

by 100 pixels, and the placement distances are 50 pixels to the right and 50 pixels up.

You see at the upper left the starting square. Next to it is an L-shaped figure made of three shrunken copies of the former picture collaged as described. Next to the L is a picture made of a collage of three shrunken L's, and so on. Finally, after a few more iterations (⋯), the wonderfully intricate shape shown in Figure 1.16 emerges. This complex of triangles-inside-triangles-inside-triangles . . . is called a **Sierpinski Gasket.**

Note that (in one way of looking at it) this Gasket is composed of three copies of itself each reduced from the whole by a factor of $\frac{1}{2}$ on each side: one in the lower left corner, a second in the lower right corner, the third in the upper left corner. Note also that when the rules we are using in this example are applied to this Gasket, the new picture produced is indistinguishable from the old. This Gasket is said to be *invariant* under the collage of the three given rules. In this sense it is "fixed," much like the stationary blob we described in the video feedback experiment. A moment's reflection suggests that the starting picture is actually irrelevant to the Gasket that finally appears. For example, in Figure 1.17 we apply the same rules to the picture of a cat. In the second picture 3 reduced cats appear; in the third, 9 cats, reduced even more; in the fourth, 27; by the fifth picture (fourth

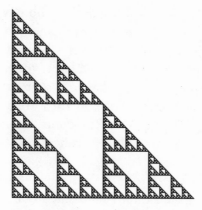

Figure 1.16   *A Sierpinski Gasket.*

iteration), 81 tiny cats begin to distribute themselves very clearly over the outline of the Gasket. Soon the cats are replaced by single pixels, and the collection is indistinguishable from that in Figure 1.16.

The Chaos Game

The next stop in our introductory stroll through the hall of complex tricks is the "Chaos Game" popularized in a variety of

Figure 1.17   *A Gasket made of cats.*

contexts by Michael Barnsley (about whom we will say more later) [62]. Perhaps you have seen the game played before. It has appeared on several TV shows and has been described in magazine articles and other books. If you have, you may remember the remarkable outcome. If you haven't seen it played or don't remember the punch line, you're in for a treat. It goes like this.

The Chaos Game (Three Corners, Two-thirds Version): You need a die (that is, one half of a pair of dice), a sheet of paper, a ruler, and a pen or pencil.

- On the sheet mark off three points not all on the same line; were they connected with lines, they would be the vertices of a triangle. Let's call these "vertices" A, B, and C. Mark also a starting point P_0.

- As the game unfolds, new points P_1, P_2, P_3, . . . will be created. (Don't worry, you aren't going to have to keep track of the names of these points.) Let faces 1 and 2 of the die correspond to vertex A, faces 3 and 4 to B, faces 5 and 6 to C. Roll the die. One of its faces will come up; suppose it's 3. Because 3 corresponds to vertex B, *imagine* drawing a line between B and the starting point P_0 and marking a new point (P_1) along the line connecting B and P_0 and two-thirds of the way from B. Don't actually draw the line; your sheet will get cluttered really fast if you do.

- Fine. Now repeat the process, using P_1 as your new starting point. Roll the die. A new face pops up. Say it's 6. That's supposed to correspond to vertex C. Imagine a line connecting C and P_1, and locate the new point, P_2, along it, two-thirds of the way from C.

- Repeat the process one more time. Say 4 comes up. P_3 is then two-thirds of the way from B (again) to P_2. Look at Figure 1.18 to see an example of how these first points might appear. We've drawn the lines in and labeled the points for the sake of clarity; *you* don't have to

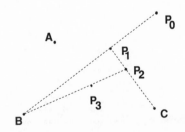

Figure 1.18   *The first few points of a typical Chaos Game.*

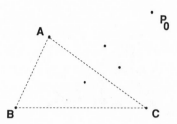

Figure 1.19   *As the Chaos Game proceeds, successive points are attracted inside the triangle formed by the lines connecting A, B, and C. Once inside, all future points stay inside.*

do that. Now repeat the process again and again and again. This is the fun part: guess what pattern the points make after you've marked off a few thousand or so.

To guide your guessing, consider Figure 1.19. This figure is the same as 1.18 except that we've removed the connecting lines and the labels P_1, P_2, and so on. Now, however, we have drawn in lines connecting A to B, B to C, and C to A for instructional purposes. The point in the lower part of the triangle between B and C is actually the last new point P_3 described above. Is it clear that once a "new" point gets inside the triangle ABC, all successive "new" points must lie inside the triangle also? (According to the rules, you have to go two-thirds the way from A, B, or C each time, so there's no way to get back out once you're inside.) What happens is that the successive points of the game keep hopping around helter-skelter (that is to say, unpredictably — as long as your die is fair and you don't bias the rolls somehow) inside the bounding triangle as the game proceeds. Does that help your intuition any?

When asked this question, many smart folks guess that eventually the triangle ABC becomes completely filled in. (Those folks realize that if the die is fair, long strings of one letter — A, A, A, A, . . ., for example — are possible (though perhaps infrequent) so that the regions near the vertices *can* get filled in.) In fact, when the Game is played long enough the triangle *does* fill in — although, curiously, not uniformly. What appears is shown in Figure 1.20.

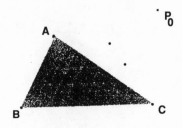

Figure 1.20   *The Chaos Game (Three Corners, Two-thirds Version).*

Well, aside from the slightly fuzzy dark criss-cross in Figure 1.20, this doesn't seem like much of a game. So, okay, let's try a very minor variation. We'll play the game exactly as before, only now, instead of plotting each new point two-thirds of the way from the newly chosen vertex to the previous point, we'll plot it one-half of the way. Figure 1.21 shows how the new construction would go. Instead of the previous points P_1, P_2, P_3, and so on, we get new points P'_1, P'_2, P'_3, and so on (the stars), assuming the same die-rolling sequence. The *primed* points are different from the *unprimed* points, but again, once the points get inside the boundary of the triangle ABC, they stay in there. So what's new? Why is this interesting? We should eventually completely fill the interior of the triangle — same as before, say the smart folks, right? This time they're wrong! What actually occurs can be seen in Figure 1.22.

Figure 1.22 shows a computer screen with a few thousand points drawn at random by the Chaos Game (Three Corners, One-half Version). It's another Gasket! Such an outcome, with its precise, lacy structure, is, at the very least, counter to our intuition; in fact, at first sight it seems impossible. The newly drawn points are supposed to be hopping around inside the

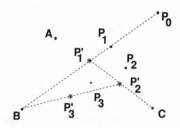

Figure 1.21   *The Chaos Game (Three Corners, One-half Version).*

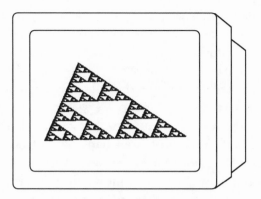

Figure 1.22   *Computer-generated three-corner Chaos Game picture.*

triangle ABC willy-nilly. Yet somehow, they don't land just any old place. Perhaps you might think we have tricked you by choosing our starting point in such a special way that this amazingly complex and detailed picture results. But that is not the case. In drawing Figure 1.22 we threw away the first few points; *if we had chosen any other starting point, we would have gotten exactly the same result after a short time.* That is, there is something robust or permanent about the picture. It really doesn't matter where we start, as long as we play the game the same way. The picture of the Gasket is determined by where you place the vertices and by the rules of the game; change the rules and you get a new picture.

In the Chaos Game, there are three important "control knobs": how far you go from the newly selected vertex to the previous point before putting down the next point, how many corners there are in the game, and the orientation of the corners to each other. We've just seen that how far you go makes a big difference when you are playing with three corners. What happens if we change the number of corners?

As an illustration, let's take four corners at the vertices of a square. We'll keep the rule "select a vertex at random, and go halfway from it toward the last point to make the next point." (This is the *Four Corners (Square), One-half Version* of the Chaos Game. You have to generate a random sequence of four values to play this version. You can use a spinner with four outcomes, for example, or, better yet, a computer that has been programmed to do such things.) What do you think you get when you've played this game for a long time? Perhaps a four-cornered variant of a Gasket, something like squares-within-squares-within-squares . . . ? Alas, the answer is merely a filled-in square. Nothing pretty. (What do you think happens when the corners are not at the vertices of a square? Try it.)

Despite its apparently mundane character, the Four Corners (Square), One-half Version of the Chaos Game has an interesting potential use. As pointed out by Jeffrey [63], it can tell us something about how DNA, the molecule that stores biological information, is put together! Here's how. DNA is composed of four chemically different building blocks, called bases, which are designated by the letters A, C, G, and T. All biological information is thought to be encoded in "words" spelled from these letters. If we look at the string of A's, C's, G's, and T's making up a given gene (a piece of DNA encoding the rules for making a protein), it's very hard for the untrained eye to see much rhyme or reason in it. Here's an example, using the first 250 or so bases in the DNA coding for the enzyme amylase:

GAATTCAAGTTTGGTGCAAAACTTGGCACAGTTATCCGCAAG
TGGAATGGAGAGAAGATGTCCTATTTAAAGTAAATATATACG
ATTTTGTCATTTGTTCTGTCATACATCTGTTGTCATTTTCTTAA
ATATTGTAACTTAAATTGTTGATTATTAGTTAGGCTTATTGTT
CATTTATCCTTAATTAATTATGTTTTTCATTTGATACATCAGT
CACCTGATAACAGCTGAAATCTAAAGTATCACTTAGTGAGTT
TTGTTGGGTTGTGTT

But if we make a square with corners labeled A, C, G, and T, take a starting point in the middle of the square, and then play the Chaos Game by going halfway to a previous point from the vertex whose letter appears next in a given DNA string (many such strings have been identified, in everything from primitive organisms to humans), some of the syntax of this code jumps out.

Figure 1.23 is the Chaos Game picture generated by the DNA for the enzyme amylase when all of the bases, not just those shown above, are used. It is *not* a filled-in square. (Incidentally, the picture that is produced by a string of DNA depends some-

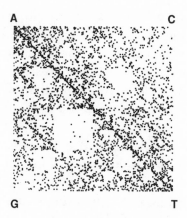

Figure 1.23   *Four-cornered Chaos Game picture generated by the DNA encoding the enzyme amylase.*

what on the order of the letters chosen for the vertices. All pictures contain the same information, however.) How to read what information is in this picture is a question we will return to in Chapter Two.

The Mandelbrot Set

In the previous sections we saw how the Sierpinski Gasket — an extraordinarily intricate latticework of triangles within triangles — can be constructed in two different ways by applying a small set of rules over and over again. It turns out that many other beautiful and exotic pictures can be drawn by similar repetitive procedures. Each picture is generated by a set of rules; each set of rules produces a unique picture.

The next exhibit in our introductory tour features a few examples of pictures (generated by repetitive application of simple rules) all associated with what has become a rather famous object called the **Mandelbrot Set.** The visual, mathematical, and metaphorical richness of this object was suggested by Mandelbrot's pioneering work [64] and has been confirmed and embellished by countless contributions from investigators around the world. None of the pictures shown in this section, nor little else of this entire book, for that matter, would exist without the aid of the modern computer and its incredible power to produce graphic images. Though many of the ideas with which Mandelbrot is closely associated had their origins decades earlier, he was the first to argue forcefully for their applicability in describing the real world, and visualization via computer graphics added untold punch to those arguments.

The Mandelbrot Set is the stuff colored black in Plate 2. We will discuss how this set is constructed in Chapter Seven. Suffice it here to say that a simple iterative process is behind it. So simple, in fact, that the rich, surprising structure seen in Plate 2 and in the related images of Plates 3 – 8 can scarcely be imagined to arise from it. For now, just enjoy the visual treats found in these pictures.

Snowflakes, Ferns, Islands, and Poons

Let's return to the Sierpinski Gasket for a moment. You can't help but be struck by the recurrent character of the structure you see in Figure 1.16. As we have noted, on the coarsest level the big triangle is composed of three smaller copies of itself. Each of these is in turn composed of three smaller copies of itself, and so on into infinite regress. (Actually, the "resolution" of the picture allows you to see only a few of these copies; you have to imagine the infinite part.) A *paper and pencil construction of the Gasket* follows this algorithm:

Draw a triangle.

Mark off the midpoints of each of the sides and connect them with straight line segments.

The four resulting triangles are all smaller copies of the original. "Remove" the middle one (in the sense that we aren't going to draw anything further inside it).

Repeat the second and third steps for each of the remaining triangles.

Repeat again and again and. . . .

Try it, it's easy. You may very well be wondering how this algorithm, the collage trick, and the Chaos Game could result in the same picture. Don't fret; it's *not* obvious. We'll come back to the connection later.

Pictures like the Gasket that are made of smaller copies of themselves at all orders of magnification are said to be *self-similar fractals*. (The word *similar* is used in its geometric sense here. Two triangles are similar, for example, if they have all the same angles. The smaller is an *undistorted,* shrunken copy of the larger.) We can make lots of other pictures with fractal-like aspects. One that is easy to make and is a pleasant shape is the *Koch Snowflake*. This is a way to make it:

Draw an equilateral triangle.

Erase the middle third of each of the sides and construct little equilateral tents over the resulting gaps.

Repeat for each side — that is, each straight line segment — that remains.

Repeat again and again and. . . .

The construction is shown in Figure 1.24. In the eventual Snowflake, every small enough segment of the perimeter contains smaller copies of itself on all scales of magnification.

Although the Koch Snowflake doesn't look exactly like any enlarged snowflake you've ever seen — because, among other things, it's too perfectly symmetrical — it does bear similarities

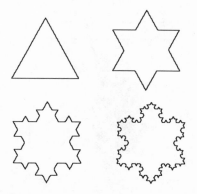

Figure 1.24 *A few steps in the construction of the Koch Snowflake.*

Figure 1.25   *A real snowflake exhibits some assymetry, especially at high magnification.*

to real snowflakes (see Figure 1.25). Perhaps we would be justified in saying that the Koch Snowflake provides us with a slightly unnatural replica of reality, at least as far as snowflakes go.

Now consider Figure 1.26, a version of a fractal image originated by Barnsley [65]. This object looks a lot like what the botanist calls a black spleenwort fern, but it really isn't one. Look closely at the small fronds that protrude on each side of the main stem. These fronds are exact, small copies of the whole fern, the ones to the right being flipped so that they curve from right to left instead of from left to right. In fact, each segment of a frond is a copy of the whole fern, and so on. This fern is too perfectly symmetrical to be one you might encounter in a real forest. Again, it is a slightly unnatural replica of reality. It, like the Sierpinski Gasket, the Mandelbrot Set, and the Koch Snowflake, was "grown" by the repetitive application of a small set of rules. (The fern takes only four rules!)

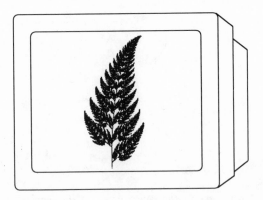

Figure 1.26   *A fractal fern.*

Real snowflakes and real ferns are not made up of exact little copies of themselves, which in turn are made up of exact little copies of themselves, which in turn. . . . Nonetheless, small bits of a snowflake or a fern may look strikingly like the whole thing. That is, many (most?) natural objects reveal a certain degree of self-similarity on a number of length scales. This approximate self-similarity is said to be *statistical* because the pieces look the same *on the average*. We differentiate between mathematical, self-similar fractals and these *natural fractals,* then, by the degree to which the parts are exactly constructed from the whole by precise acts of shrinking, rotating, and squashing. In mathematical fractals the parts are copies of the whole transformed, typically, by a small number of relatively simple operations; in natural fractals the parts are only reminiscent of the whole.

One of the central characteristics of fractal objects is their (at least approximate) invariance under magnification. Because the pieces of the object are shrunken copies of the whole, it is often difficult to tell at what scale you are viewing the object. The coastline of an island in a lake is a good example. When viewed from an airplane it appears jagged, irregular, with all kinds of protuberances and indentations. Perhaps there are a few large, rocky outcroppings immediately off-shore. At a lower altitude, boulders on the shoreline begin to be resolved; in the statistical sense we introduced above, these look essentially like the island itself. Rocks in the water near these boulders look much like the island's outcroppings. At even closer scale, the rocks become islands with nearby pebbles the outcroppings. And so on. It might even be that the island we started with is itself just offshore of an island continent in an ocean of the Earth. Without some obvious marker whose length is well known — such as a boat (careful, boats come in lots of different sizes, too) or a swimmer — it is difficult to tell whether we are looking at an island or a boulder or a continent.

Mathematical fractals exhibit scale ambiguity at all levels of magnification (in the mind's eye), but any physical fractal or physical representation of a mathematical fractal (such as a picture on a computer screen) has a largest scale and a smallest scale beyond which the character of the object changes. Molecules, after all, are *not* shrunken copies of the bodies they make up.

Nonetheless, distance confusion often occurs when we view natural objects. (Just how far away is that mountain or that cloud bank, anyway?)

Figures 1.27–1.33 are photographs of natural objects that illustrate the fact that self-similarity leads to ambiguity of scale.

Because fractals capture some of the essence of real objects and because visual art often attempts to do a similar thing, it is not surprising that some art employs fractal properties. Though the formal, mathematical ideas of fractals have been well known for only a few years, several artists have hit on the esthetic value of self-similarity without help from the mathematicians (see Figure 1.33). And consider Plate 9, which shows a rather typical later painting by the contemporary American artist Larry Poons ("Rum Boat," 1983). Richly textured with streaks and knots, this work is a showcase of self-similarity. As the art historian and critic Daniel Robbins says [66],

(Poons) has created a new kind of abstract art of globular intricacy, sensual power, and intellectual challenge. . . . He has elevated the relationship between order and inexactitude to a level comparable

Figure 1.27 *A blossom of Queen Anne's Lace and its many similar sub-blossoms.*

Figure 1.28   *Is this a satellite photo of cloud formations off the coast of Australia, an aerial shot of the surf breaking near Monterey, or turbulent gurgles in a Jacuzzi taken from a height of a few inches?*

Figure 1.29   *From what distance was this photo of a barren escarpment taken?*

Figure 1.30   *This vine network may suggest to you a certain range of sizes, up to inches, perhaps. But the roots of the strangler fig tree form an essentially identical network, and the largest root can be as thick as the girth of a sturdy human.*

Figure 1.31   *This wonderful tangle of branches, stems, and twigs shows a remarkable self-similarity. Is it obvious that it is not a photomicrograph of a capillary system?*

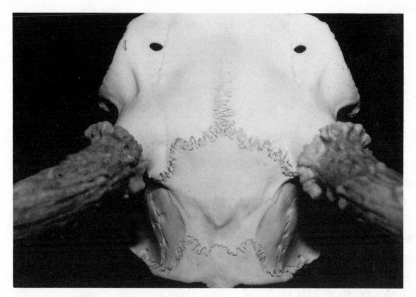

Figure 1.32   *Self-similar anatomical structures are commonplace. An example is the interwoven sutures in the skull of a mule deer.*

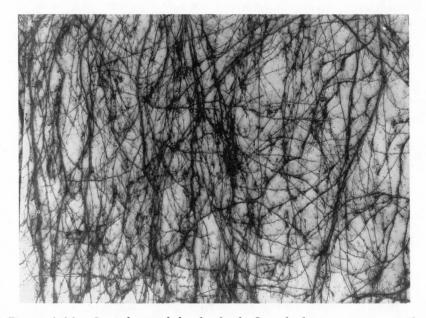

Figure 1.33   *Does the gnarled web of a leafless climbing ivy conjure up the drip paintings of Jackson Pollock?*

with the parallel discoveries of fractal geometry; but Poons is making art, not geometry. He seeks for beauty and large ideas, but in the unmeasurable realm of the painting's effect, not for a fractal dimension.

To Noise and Bach

A common element of an electrical circuit is a resistor, a material that, while permitting the flow of electricity through it, limits the size of that current. If we connect a resistor to the terminals of a battery, a current flows through the resistor. This current is observed to be *approximately* constant (at least until the battery begins to "run down"). But when we examine the current with great care (with some fancy equipment), we find that it actually undergoes very slight fluctuations around the "constant" value, as shown in Figure 1.34.

If you set your stereo to "phono" or "tape" and turn up the speaker volume, you will hear a hiss. This hiss is produced by the fluctuations in current through the electronic elements in your stereo's circuitry. Because of the hissy sound associated with them, these current fluctuations are said to constitute *noise*. The "noisy" electrical current of Figure 1.34 looks pretty much the same no matter how long a piece of it we examine: it doesn't have a clear-cut characteristic time scale. Because fractals in space — like those in the previous section — have no characteristic length scale, it is useful to think of such erratic, fluctuating signals as *fractals in time*.

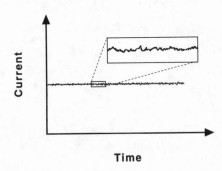

Figure 1.34 *An approximately constant current actually fluctuates slightly.*

Noisy signals include not only the current fluctuations we have just depicted but also many other interesting phenomena. A few examples include price variations of stocks, the irregular fibrillation of a malfunctioning heart, the electrical activity of the brain, the clumping of cars on an expressway, and the extent of the annual floods of the Nile. Even spatial patterns can be thought of as "noisy": altitude variations along a mountain range, the distribution of gold deposits over the face of the Earth, and the occurrence of A's, C's, G's, and T's along a strand of DNA are some of the many irregular spatial signals that surround us. Noise can be characterized by its jaggedness. Some noisy signals are so irregular that even when a long history of past results is available as a guide, we have no chance of predicting what outcomes will happen next. Other irregular signals are more orderly; detailed prediction may still be elusive for these signals, but some bounds can often be placed on next outcomes. The former kind of noise is said to lack *correlation;* the outcomes from instant to instant are completely independent. In less jagged noises the outcomes are somewhat correlated — in other words, if this and this happened before, then only such and so can happen next (though we may not know which of several outcomes that belong to "such and so" will actually occur). Understanding the character of irregular signals is a central issue in the study of complexity and is a fundamental theme of this book. We will examine noise in much more detail later.

While studying the nature and causes of correlated noise in electronic devices for his doctoral thesis work in the early 1970s, Richard Voss made a remarkable discovery [67]. During a lull in data collection, Voss fed the signal from a local classical music station into the same equipment he was using to analyze the small fluctuations of current in his circuits. Because classical music consists of intricate patterns of tones, rhythms, durations, and emphases, Voss reasoned, a statistical analysis of its variations should reveal complexity and might even be instructive to examine. Astonishingly, the statistical treatment of classical music Voss performed turned out to look very similar to that of the noise he was studying in his experiments. So, for that matter, did the analyses of country and western music, rock and roll, and jazz. Inspired by this apparent kinship between most forms of music and the noise of electronic elements, Voss wondered about the reciprocity of the relationship: could real noise be used to make real music? In fact, with a judicious choice of rules for note selection, he was able to convert current fluctuations into quite reasonably sounding tunes—almost music but, like the fern and the snowflake of the preceding section, not *quite* the genuine article.

In subsequent studies, Voss and others have demonstrated that how close the tone patterns generated by noise are to real music depends on the correlations in the noise source. Noise with no correlation produces a haphazard, irritating jumble of notes (not unlike the random compositions of John Cage). Noise with too much correlation contains little surprise; it sounds like someone playing scales. But noise with just the right amount of correlation produces tone combinations that exhibit an interesting mixture of coherence and novelty.

If the correlation content of noise can be sensed through its translation into audible tones, why not try visible colors? In the 1950s, the American artist Ellsworth Kelly produced a series of paintings that were intended to be devoid of esthetic choice. These paintings are rectangular grids of small squares of color, the placement of each being drawn by chance (roughly akin to the aforementioned chance music of Cage). Plate 10 shows an example; it is very similar to many of the Kelly paintings. This "painting" was generated by selecting spectral colors, plus black and white, using an uncorrelated noise source. You can see that the resulting colors are juxtaposed without any apparent guiding theme. Plate 11, on the other hand, shows what a highly correlated noise source produces: reasonably rigidly structured bands. If you "read" the colors left to right, row after row, you see gradual excursions "up" the spectrum (red-orange-yellow-green-blue-violet) followed by gradual descents (punctuated by an occasional black—infrared—or white—ultraviolet—square). This pattern is analogous to the "scale playing" noted in the previous paragraph. Finally, in Plate 12, you see the results of using noise with intermediate correlation: not so erratic as the first result, not so structured as the second. Which of these "paintings" do you find most pleasing?

Life

The last stop we make in our introductory tour involves cellular automata. A *cellular automaton* "lives" in a computer; it consists of a set of rules that determines whether a "cell" is dead or alive in every succeeding generation. A given population of the automaton is determined by its starting configuration and by the number of generations that have passed since the start.

Here's a specific, simple, one-dimensional example. The population lives in an infinitely long, linear row of possible slots, one cell per slot. A slot either is occupied by a living cell or is not, depending only on itself and on its right and left neighbors in the immediate past generation. If in that generation it was *not* occupied and, of its two neighbors, only the slot to the right was, or if in that generation it was *not* occupied and, of its two neighbors, only the slot to the left was, then that slot is occupied in the current generation. Otherwise, the slot is unoccupied.

Of course, these rules don't correspond to how any real cell populations behave. They are so simple you can figure out how a population would change from generation to generation by hand. Draw a string of seven empty boxes imagined as a portion of a long string of empty boxes. Designate that the middle box (box 4) is occupied with a living cell by putting an X in it. Now draw seven boxes, indicating where cells are alive in the next generation. Repeat one more time. (*Answers:* In the second generation, boxes 3 and 5 are occupied. In the third generation, boxes 2, 4, and 6 are occupied.)

Figure 1.35 shows how a population governed by these rules evolves if a random initial occupation of the available slots is given. In this picture each horizontal line is made up of white or black dots representing, respectively, empty or occupied slots. Each successive line starting from the top and working downward represents a new generation (that is, time increases in the downward direction). You can see that this time-lapse photograph is incredibly intricate. The empty triangular gaps conjure up an eerie echo of things we've seen before. Sure enough, if we start with a single cell alive in the first generation, we get Figure 1.36!

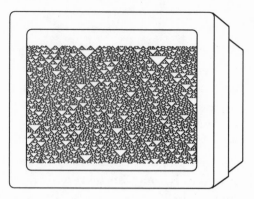

Figure 1.35 *A one-dimensional cellular automaton population growing out of a random initial state.*

Slight alterations of these rules can lead to amazing patterns in the time-lapse photographs of one-dimensional automata. In Figure 1.37, for example, starting from a single cell we get an amazing juxtaposition of patterned symmetry to the left and a rubble heap of disorder to the right as time goes on.

For cellular automata living in two-dimensions—that is, populating slots spread out over a plane rather than along a line—even more interesting and suggestive behavior can be observed. For example, in Figure 1.38 we track the evolution of a two-dimensional population in which the rules involve a given

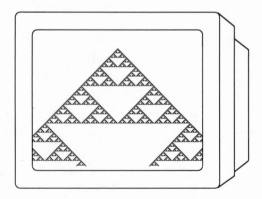

Figure 1.36 *A one-dimensional cellular automaton population growing out of a single cell.*

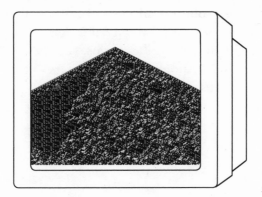

Figure 1.37   *Another one-dimensional cellular automaton population growing out of a single cell.*

slot and its eight neighbors. (Imagine a little three-by-three grid with the center as the slot in question. Cells living in the eight surrounding slots determine the future occupation of the center slot in this automaton.) In the first generation (a), the available slots are randomly occupied. (A stew of scrambled atoms?) After

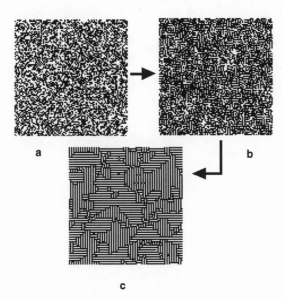

Figure 1.38   *A two-dimensional cellular automaton. In a, the starting state. In b, the state after 10 generations. In c, the state after a few thousand generations.*

10 generations (b), some primitive wormlike structures begin to emerge. (Large molecules?) Much later (c), these worms have sorted out into highly ordered domains. (Colonies of amebas?) In this automaton, order spontaneously appears out of the primordial pool of utter disarray.

Neat tricks, all.
It's time to start figuring out how they work.

2

Only God Can Make a Tree?

THIS CHAPTER DEALS with a potentially important practical use of fractal geometry: making pictures by computer in a way that requires incredibly less memory than any method yet devised. Because memory is money in the information business, this topic is much more than just intellectually interesting (though it is certainly that). Still in its infancy, the **iterated function system** method is attributed in its mathematical form to John Hutchinson [68] and was popularized and implemented for image processing by Michael Barnsley [69]. We cannot yet say what new products and technologies it will spawn, but some have appeared already. In the next century, electronically transmitted, visual information may rely heavily on the principles we describe below. A side benefit of learning about iterated function systems is that we may well also learn something about how information is stored in living cells — a matter of fundamental and clinical importance that is only partially understood at this time.

The Gasket Shall Be Our Guide

Look again at Figure 1.16, the Sierpinski Gasket. The Gasket consists of a collection of smaller copies of itself. There are three main smaller copies, contained within the shaded triangular regions shown in Figure 2.1.

Our task is to try to tell a computer exactly how to make these three copies and paste them together. Computers deal poorly

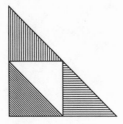

Figure 2.1   *The three main regions of a Sierpinski Gasket.*

with such qualitative, verbal instructions as "make a copy half as big as the previous picture; make a second identical copy and place it directly to the right; make a third identical copy and place it the same distance directly above." To instruct the computer properly, we need to be more exact. It helps to refer to the pixels that make up the computer's display. We choose a reference pixel as the "origin." Then we label each pixel by a set of coordinates (x = some number of pixels away from the origin in the horizontal direction, y = some number of pixels away from the origin in the vertical direction). In this scheme x is a positive (negative) number if the designated pixel is to the right (left) of the origin, and y is positive (negative) if the pixel is up (down) from the origin. A "picture" then corresponds to a set of (x, y) values — that is, the pixels that are lit. (Note that if your monitor is set with a white background — that is, if you are making a black-on-white picture — the "lit" pixels will be black.)

Using this notation we can tell the computer how to make a new picture that is half as big (on each side) as the previous one. For every pixel with coordinates (x, y) that is lit on the previous picture, light one with coordinates (0.5·x, 0.5·y) on the new picture. (Computers don't use fractions; they use decimal equivalents.) Thus, for example, if the pixel at coordinates (20, 40) is lit on the old picture, the pixel at (10, 20) will be lit on the new. This rule — abbreviated "scale by a factor of 0.5" or, even more briefly, "scale by 0.5" — shrinks every point on the old picture halfway toward the fixed point located at the origin, (0, 0). This is shown in Figure 2.2. Note also that the figure indicates that every point on the smaller striped triangle comes from a point on the larger triangle through application of the rule "scale by 0.5."

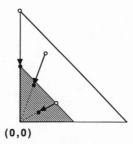

Figure 2.2   *Results of the rule "Scale by a factor of 0.5." Every point in the larger triangle is sent halfway toward (0, 0).*

How do we create shrunken pictures placed where we want them? We first scale down toward the origin and then translate the resulting picture by the desired amount. A *translation* is a motion in which all points on the picture travel along parallel paths; there is no "tumbling" as the picture moves. A translation in the x-direction is equivalent to adding (or subtracting) the same amount to (or from) the x-coordinate of every point on the picture. Scalings are determined by what fraction of the original picture the smaller picture is, regardless of the size of the original picture. For example, "scale by 0.5" takes a square 100 pixels by 100 pixels to a square 50 pixels by 50 pixels, and it takes a square 150 pixels by 150 pixels to a square 75 pixels by 75 pixels. To treat translation in the same way, we need to set a unit length and express the translation as a (possibly fractional) multiple of this unit length. For example, if the unit length is 100, then the rule "translate by +0.5 in the x-direction" takes the pixel (0, 100) to (50, 100) and the pixel (100, 0) to (150, 0). Similarly, "translate by −0.5 in the y-direction" takes (0, 100) to (0, 50) and (100, 0) to (100, −50). When we sketch by hand the effect of these translations, it is usually more convenient to deal not with the pixel numbers (which the computer uses), but with the fraction of the unit length. In these terms, the rule "translate by +0.5 in the x-direction" takes the point (0, 1) to (0.5, 1) and the point (1, 0) to (1.5, 0). Similarly, "translate by −0.5 in the y-direction" takes (0, 1) to (0, 0.5) and (1, 0) to (1, −0.5). As an example, Figure 2.3 shows how the three main regions of the Gasket of Figure 1.16 are made by scaling and translating. The

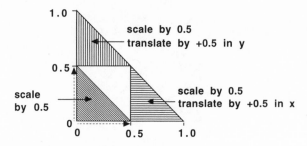

Figure 2.3 *The three main pieces of a Gasket and the rules to make them from a larger triangle.*

shadings are added later so that we can more easily distinguish the pieces.

When applied iteratively a sufficient number of times, the set of rules

Scale by 0.5.

Scale by 0.5; then translate by +0.5 in the x-direction.

Scale by 0.5; then translate by +0.5 in the y-direction.

Then collage ("glue") the pieces.

generates a Gasket like the one in Figure 1.16; the lengths of its horizontal and vertical sides are both equal to 1. As we suggested in the first chapter, it doesn't matter what picture we start with. Once formed, that Gasket is invariant under the application of these rules.

Suppose that instead we wanted to make an *equilateral* Gasket with side length equal to 1. What rules would we need then? Figure 2.4 is our guide. To make an equilateral Gasket, keep the first two rules above, but change the third so that it places a piece in just the right spot—a little to the right and a little up. "Just right" can be deduced by noting where the lower lefthand corner of the triangle formed by "scale by 0.5" (diagonal stripes) has to go in order to be the lower lefthand corner of the small triangle (vertical stripes) in the upper middle part of the larger triangle. (The x-part is easy: go over half of the base of the diagonally striped triangle—0.25. The y-part corresponds to the height of the diagonally striped triangle. Pythagoras's Theo-

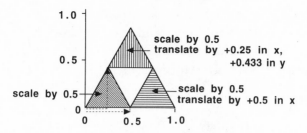

Figure 2.4   *The three main pieces of an equilateral Gasket and the rules to make them.*

rem requires that $(0.5)^2 = (0.25)^2 + y^2$, which leads to $y = 0.433. \ldots$) Of course, we don't have to be confined to nicely symmetrical triangular Gaskets. Three rules, each containing "scale by 0.5" along with appropriate translations, allow us to make Gaskets of any triangular shape we please.

We claimed at the outset of this chapter that fractal image construction held tremendous potential for practical application. Here's why. If someone asked you to light up pixels on a computer display to render an image of a Gasket, a brute-force way to do it would be to write a data file containing the coordinates of each pixel to be lit and then feed that file to a program that would do the lighting. Given that a typical computer display contains about 250,000 pixels, such a data file would be very large indeed. A much less memory-hungry method would be to write a short program consisting of the three scaling and translating rules we've been discussing here. Whereas the former method occupies thousands of lines of computer code, the latter occupies tens. The data requirements for making a Gasket on a computer monitor can be enormously reduced by taking advantage of the fact that the Gasket is made of shrunken copies of itself. And as we will soon see, such amazing data compressions are not restricted to pictures of Gaskets.

Making Some Other Pictures

Creating computer images by using fractal geometry would not be very interesting if Gaskets were the only objects that could be produced. To make other kinds of pictures, we need to generalize the picture-mangling rules we wrote down above. We don't have to stick to three rules, and we can use other kinds of transformations besides "scale by 0.5" and translate.

For example, suppose we use four rules:

Scale by 0.5.

Scale by 0.5; then translate by +0.5 in the x-direction.

Scale by 0.5; then translate by +0.5 in the x-direction and by +0.5 in the y-direction.

Scale by 0.5; then translate by +0.5 in the y-direction.

Then, finally, collage together.

Can you guess what picture emerges from the iterative implementation of this set of manglings? Figure 2.5 shows what happens under these rules to a unit square—a square with a length of 1 per side. The portion of the figure to the right has the transformed pieces labeled with the rules that made them. After gluing the pieces together, we see that under these transformations, a filled-in square becomes a filled-in square. (When we glue, the interior cross made by small square boundaries gets blurred out.) The square is invariant under the four rules listed. It (the square) is the picture that results from many iterations of the rules. (You could have started with a picture of your favorite cat or a dollar bill; a square results from every start.)

As another example, let's take three rules again but let's scale by $\frac{1}{3}$ (0.333 . . .) instead of by $\frac{1}{2}$.

Scale by 0.333.

Scale by 0.333; then translate by +0.5 in the x-direction.

Scale by 0.333; then translate by +0.5 in the y-direction.

Collage the pieces.

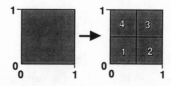

Figure 2.5   *A square is a square.*

Before reading further, try to guess what picture results from successive iteration this time.

Remember that scaling by 0.5 in the three transformation rules leads ultimately to a Gasket whose outside boundary is an isosceles right triangle. To see what will happen with the new rules, start by taking a picture of an isosceles right triangle — one with vertices at (0, 0), (1, 0), and (0, 1), say. (We could start with *any* picture; this one just happens to be convenient.) Apply the three rules to this picture. In the first generation we get three smaller triangles, but now the small triangles don't touch because they are only one-third as big as the starting triangle. Another iteration leads to nine even smaller triangles — all also not touching — and another to 27, and so on. Ultimately the ever-reducing triangles contract so far that we see them only as single pixels on our computer display. This situation is depicted in Figure 2.6.

The ultimate picture is a collection of disconnected points. Note that the little triangular smudges in the figure are really very small copies of the whole figure — themselves collections of disconnected copies on even smaller scale. Mandelbrot has introduced a term for such a collection of points, one that conjures up a graphic image: he calls them a dust of points. (You may have noticed that in each successive iteration, the total size of the picture appears to shrink a bit. That's true; in fact, the base and height of the final Gasket dust are both ¾. Do you see why?)

The dust of points in Figure 2.6 is a variant of what is called a *Cantor Set*. The prototypical Cantor Set — the so-called "middle thirds" set introduced by Georg Cantor in 1883 — is constructed with pencil and paper as follows:

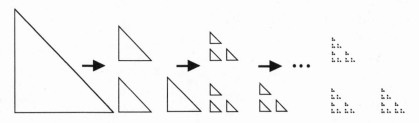

Figure 2.6   *A triangular dust.*

```
0  ————————————————————
1  ————  ————          ————  ————
2  ——  ——    ——  ——      ——  ——    ——  ——
3  -- --   -- --        -- --   -- --
4  -- --  -- --         -- --  -- --
```

Figure 2.7   *Successive iterations leading to the Cantor Middle Thirds Set.*

Take a line of length 1; erase the middle third of the line.

For every line that remains, erase the middle third.

Repeat the previous step again and again and again. . . .

Successive iterations of this construction are shown in Figure 2.7. After the fourth iteration it gets hard to discern any difference between successive pictures.

How would you construct the Cantor Middle Thirds Set (Cantor MTS) from scalings and translations? As a guide to your thinking, notice how many copies of the original line appear in the first-generation picture: there are two copies, each shrunk by a factor of 0.333. Thus the scaling part is easy. Now, what translations are necessary? To help on this point, consider Figure 2.8. For the sake of specificity, we start with a line spanning the interval $x = 0$ to $x = 1$. We mark one end of the line with a star. In the first iteration two new lines appear, one obtained by scaling the whole starting line by 0.333 toward the origin, the other obtained by scaling the whole starting line by 0.333 toward the origin and then translating that result by $\frac{2}{3} = 0.667$ in the positive x-direction. (See where the star went?) The resulting picture is essentially a line of length 1 *except* that its middle third has been removed. Can you see that applying exactly the same two transformations to the first-generation picture produces the second-generation picture shown in Figure 2.7 (and, successively,

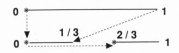

Figure 2.8   *Transformation rules for the Cantor MTS.*

the next and the next and . . .)? We can say, then, that the set of rules

Scale by 0.333 toward the origin.

Scale by 0.333 toward the origin; then translate by +0.667 in the x-direction.

Collage.

is the Cantor MTS.

Note that the Gasket was made by successively gluing together three smaller copies of each previous picture. Three transformations were required to generate it. The filled-in square was made by gluing four smaller copies together. Four transformations were required to generate the square. The Cantor Middle Thirds Set is made of a succession of two smaller copies of each previous picture; two transformations are necessary to generate it. As a general principle, *if the picture you wish to make consists of N smaller copies of itself, you will need N transformations to generate it.*

But, you might argue, not only is the Gasket of Figure 1.16 made up of 3 smaller copies of the whole, but there are also 9 even smaller copies, and 27 copies smaller than these, and 81 copies even smaller than these, and. . . . Well, by the same principle, the Cantor MTS is composed of 2, 4, 8, . . . successively smaller copies. All true! In fact, you *could* find 27 transformations that would generate the Gasket or 16 transformations that would generate the Cantor MTS. But the smallest N in each case leads to the smallest program you would have to write in order to generate the desired picture. That is, the smallest N corresponds to the optimal algorithm — optimal in the sense that it is associated with the smallest number of lines of code — for making the fractal image you wish.

Deterministic Iterated Function Systems

The pictures of the previous section are created by a subset of a much broader class of image-processing methods called *deterministic iterated function systems* (deterministic IFS). In the IFS picture-making business, a picture is a set of coordinates of lit pixels. An IFS is a collection of coordinate transformations. When an IFS is applied to a first picture (a first set of coordinates), the result is a second picture (a transformed set of coordinates). For our purposes, an IFS will permit scaling transformations (scalings), translations, rotations, and reflections. We have some practice with the first two. Let's look at what we can produce with rotations and reflections.

Back in the first chapter we talked about the Koch Snowflake. One third of the edge of the Snowflake is a self-similar fractal called a *Koch Curve*. We can make one with paper and pencil as follows: Start with a straight line segment stretching along the x-axis between 0 and 1. As in the Cantor Middle Thirds construction, remove the middle third. Now, however, instead of repeating over and over—a process that eventually yields a dust of points—erect over the central gap an equilateral tent, as shown in Figure 2.9. For every straight line segment appearing in the first generation (the second picture in the figure), repeat the process of removing the middle third and erecting an equilateral tent over the resulting gap. Repeat again and again. . . . Eventually, Figure 2.10 emerges.

Conversion of this crude geometrical construction into coordinate transformations so the curve can be drawn by a computer can be accomplished by recognizing that the Koch Curve of Figure 2.10 is built up of four smaller copies of itself, as the first-generation picture in Figure 2.9 suggests. In Figure 2.11 we redraw Figure 2.10 with the four smaller copies set out in boxes. The first copy, 1, is made by scaling the whole picture by 0.333 toward the origin. A second copy, 2, is made by scaling the whole picture by 0.333 toward the origin and then translating the result by 0.667 in the positive x-direction. The remaining two copies

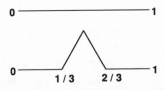

Figure 2.9　*The first step in making a Koch Curve.*

Figure 2.10   *A Koch Curve.*

are also made by first scaling the whole picture by 0.333 toward the origin, but each undergoes *rotation* as well as translation. Copy 3 is rotated about the origin (one end of the shrunken copy) by 60° in an upward manner relative to the x-axis (that is, counterclockwise) and then translated by 0.333 in the positive x-direction, as shown in Figure 2.12.

You might at first think that copy 4 is obtained by rotating about the origin by 120° and then translating by 0.667 in the positive x-direction. But that isn't right. If you did rotate copy 1 by 120°, the little hump would be pointing in the wrong direction. Try it. To get the correct orientation, copy 4 is rotated by 60° down from the x-axis (clockwise) and then translated by 0.5 in the positive x-direction (so that its upper tip is at the midpoint of the interval between 0 and 1) and by $\sqrt{3}/6 = 0.289$ in the positive y-direction (to raise its lower tip to be right on the x-axis). (See Figure 2.13.) The awkward number 0.289 results from Pythagoras's Theorem. (You should be able to figure out how.)

Finally, we consider reflections. A reflection is what a mirror does: it converts a right hand into a left. This is shown in Figure 2.14. The right hand in the right portion of the figure has its palm pointing outward. The image produced by a reflection through a plane perpendicular to the plane of the page (represented by the vertical line) is a left hand. You cannot produce

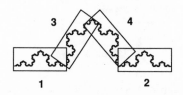

Figure 2.11   *Four main pieces of a Koch Curve.*

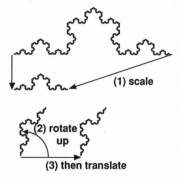

Figure 2.12   *How copy 3 gets there.*

such an image by a rotation. If you rotated the right hand 180° around an axis parallel to the line in the picture, all fingers would be pointing in the proper directions, but we would see the back of the hand instead of the palm. Every point in the reflected image originates from a point in the original; the x-coordinate of the reflected point is the negative of the x-coordinate of the point in the original, as long as we reflect across the y-axis. (The y-coordinates are the same.) Figure 2.15 shows a reflection across the x-axis. The y-coordinate of every reflected point is the negative of the point it originates from. (The x-coordinates are the same.)

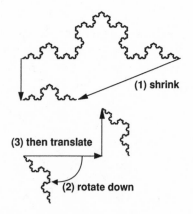

Figure 2.13   *How copy 4 gets there.*

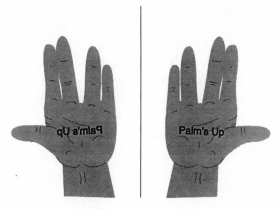

Figure 2.14 *Reflection across the y-axis.*

Figure 2.15 *Reflection across the x-axis.*

Random Iterated Function Systems

Figure 2.16 shows a sequence of images (it is read from left to right) produced by the deterministic IFS algorithm for the Koch Curve. It starts with the "face" in the upper lefthand panel. Much in the same manner in which the cat was transmogrified into the Gasket in Figure 1.17, each successive panel shows disappearing faces and an ever closer approximation to the final image (Figure 2.10). This way of constructing the Koch Curve is memory-intensive. Each panel requires storage of the pixels that are lit in that panel *and* the pixels that were lit in the previous panel. Two whole pictures have to be stored each time a new picture is made. Because so much memory is required, the process is also fairly slow. Thus, before going on to describe how the transformations we have outlined can be used to make images of real objects (mountains, trees, flowers, and even human faces), we digress briefly to describe a much faster, much less memory-intensive IFS algorithm — one that is actually used in practice.

You may have wondered how the Chaos Game played in Chapter One and the IFS transformations — scale and translate — can both make very nice pictures of Gaskets. If we play the Chaos Game with the vertices of an isosceles right triangle, selecting vertices one after the other randomly and each time placing a new dot (lighting a new pixel) halfway from the vertex selected to the previous dot, Figure 1.16 eventually emerges (plus perhaps

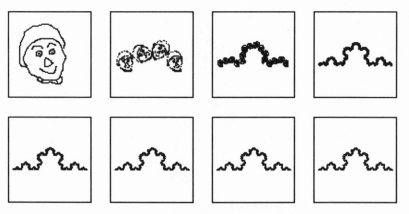

Figure 2.16   *A face is transformed into a Koch Curve.*

a few stray dots coming from the first few iterations). Similarly, if we use the IFS rules "scale by 0.5, scale by 0.5 and translate +0.5 in x, scale by 0.5 and translate +0.5 in y" on any starting picture, we also get Figure 1.16.

To see why these two seemingly unrelated processes yield the same result, consider Figure 2.17. In the background is a ghost of a Gasket. Point A is on the Gasket. Under the rule "scale by 0.5 [toward 0]," A is transformed into a second point, A′, also on the Gasket — one on a line halfway between A and 0. Point B is not on the Gasket. Under the same rule, B goes to a point, B′, on a line halfway between B and 0; that point, as you can see, is closer to being on the Gasket than was B. Then, under the rule "scale by 0.5 [toward 0], translate +0.5 in y," A′ is taken to A″ (halfway from A′ to 1) and B′ is taken to B″ (halfway from B′ to 1). A″, like its predecessors A and A′, is still exactly on the Gasket; B″ is even closer to it than was B′.

The rules of the Chaos Game are IFS rules selected randomly one at a time. Successive points generated by the Chaos Game get closer and closer to actually lying on the Gasket. Eventually, the accumulated set of such points traces out a fairly decent representation of the Gasket coded by the underlying rules. It should be clear from this discussion that we could take *any* set of IFS rules — not just Gasket rules — and light up pixels sequentially by transforming the coordinates of the last pixel lit into the coordinates of the next pixel by selecting randomly from the set. The IFS rules always refer to a fixed origin, so we could even start by lighting up the pixel there as the starting point. If that point is on the picture that the rules code, all successive pixels

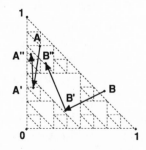

Figure 2.17   *Points near the ghost of a Gasket.*

will also be on the picture; thus there are no wasted points. The picture so produced emerges point by point like a pointillist painting. Because we need only the coordinates of the last pixel each time to manufacture the coordinates of the next, a savings of a factor of 2 can be achieved in the memory requirements of the construction. And because only one point is transformed in each update (rather than the whole set of points, as in the deterministic method), a tremendous time savings is available as well. Consequently, all practical applications of IFS image construction rely on this "random rule selection" trick.

Pictures from Nature

The Gasket, the Cantor Set, and the Koch Curve are all highly symmetrical mathematical fractals; simple geometrical transformations, such as rotations or reflections in a mirror, leave them unchanged in appearance. In nature, on the other hand, we see many objects that display degrees of self-similarity but are not exactly symmetrical. Mountain ranges and coastlines and trees do not typically remain invariant under simple geometrical transformations. One part of a mountain range is very much like another part, but there are distortions — scalings or stretchings of different sizes in different directions. To render faithful images of the things of nature, we need to generalize the transformations described previously.

We need distortion-producing transformations: scale by a certain amount in the x-direction and by a different amount in the y-direction; rotate all lines parallel to the x-axis by one angle and all lines parallel to the y-axis by another. An IFS rule can be thought of as a set of parameters that we call R, S, Theta, Phi, E, F, and Prob. The parameters R and S correspond to scaling factors along the x- and y-directions, respectively. Suppose, for example, that in one IFS rule we have R = 0.5 and S = 0.333. Then this part of the rule takes a point at (0.6, 2.0) to the point $([0.5] \cdot [0.6], [0.333] \cdot [2.0]) = (0.3, 0.666)$. The distorting effect of different scaling factors is displayed in Figure 2.18.

The parameters Theta and Phi produce rotations, the first relative to the x-axis, the second relative to the y-axis. To see

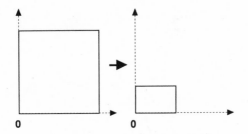

Figure 2.18   *A square shrunken by a factor of 0.500 in the x-direction, and by a factor of 0.333 in the y-direction.*

what these transformations do, consider Figure 2.19. In this figure, Theta is −30°, "−" signifying clockwise, and Phi is +15°, + signifying counterclockwise. All rotations are around the fixed origin. If Theta = Phi, the equal rotation leads to a reorientation but to no distortion. When Theta does not equal Phi, the effect is equivalent to a kind of plastic distortion.

Parameters E and F represent translations in the x- and y-directions, respectively. We've already seen how those work. Note that order is important in the application of each rule: scaling is done first, rotating second, and translating last. Doing the parts of the rule in a different order produces a different result.

Finally, Prob stands for probability. In the random IFS algorithm, we can assign different likelihoods for each rule being

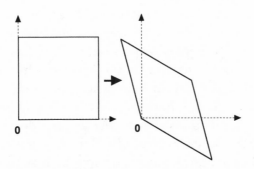

Figure 2.19   *A square with horizontal parts rotated by 30° in the clockwise sense (Theta) and vertical parts rotated by 15° in the counterclockwise sense (Phi).*

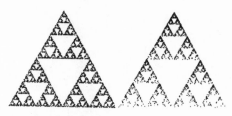

Figure 2.20   *Two Gaskets, the lefthand one done with equal probabilities, the righthand one with unequal probabilities.*

chosen. The result is to generate unequal darkening of the picture. This is demonstrated in Figure 2.20. The difference between the Gaskets in the figure is that on the left, each of the three Gasket rules was chosen with the same frequency, whereas on the right, the rule making the upper, middle portion was chosen 60% of the time and the other two rules 20% of the time each. Note that the upper, middle portion is darker than the other portions, and note that the upper, middle of every subsection is darker than the rest of that subsection, as well.

These generalized coordinate transformations can squeeze and distort a picture by pushing and pulling it differently in different directions. They have the fancy name *affine transformations*. These transformations take a "picture"—a bounded (that is, occupying a finite region of space) set of points—into another picture. Repeated application of affine transformations that are *contractive* (including scaling) produces a sequence of pictures that converges to an unchanging set of points—the picture that corresponds to the rules. We have talked only about two-dimensional pictures—that is, pictures existing in a plane, such as a computer display screen or a sheet of paper. But one can imagine the same kinds of transformations on sets of coordinates in three or higher dimensions (leading to a kind of generalized hologram)!

Incidentally, you should be aware that in the making of images through the use of iterated transformations, identical outcomes can result from different sets of rules. Any given set of rules codes a unique picture, but a given picture is not coded by a unique set of rules. One example of this statement is the realization that any fractal can be decomposed into larger and larger

Figure 2.21   *A tree grown by IFS rules.*

numbers of smaller and smaller copies; each set of copies corresponds to a different set of rules. But combinations of transformations often produce identical results. For example, an empty equilateral triangle (one whose back and front are indistinguishable) with a vertex at the origin can be transformed into itself by first rotating it around the origin by 120° and then reflecting across the y-axis. (Try it.)

Let's address the issue of creating natural pictures. Figure 2.21 shows a pretty reasonable facsimile of a tree grown by the random IFS construction. The table below contains the IFS parameters from which this tree was grown.

The first thing that strikes you when looking at this table is that there are only six rules. Then you see that the rules required are not obvious. You can also see a bit of the full generality of the transformations: there are scaling distortions, reflections, and rotations left and right, and the tree isn't filled in uniformly. The

No.	R	S	Theta	Phi	E	F	Prob
1	0.050	0.600	0.000	0.000	0.000	0.000	0.100
2	0.050	−0.500	0.000	0.000	0.000	1.000	0.100
3	0.600	0.500	40.000	40.000	0.000	0.600	0.200
4	0.500	0.450	20.000	20.000	0.000	1.100	0.200
5	0.500	0.550	−30.000	−30.000	0.000	1.000	0.200
6	0.550	0.400	−40.000	−40.000	0.000	0.700	0.200

table says that the tree is a fractal with six main pieces. To see what those pieces look like, we can do the following:

Start at the origin.

Pick the six rules one at a time, randomly (with rules 3–6 picked twice as frequently as rules 1 and 2).

Transform the coordinates of each pixel according to the rule chosen. Light the new pixel only if the rule producing it is a 1.

If we light only those pixels where 1 was the last rule chosen, we construct a picture of what rule 1 does to the whole tree. Similarly, if we rerun the procedure but light only those pixels where 2 was the last rule chosen, we construct a picture of what happens to the whole tree under transformation 2, and so on. The results of such selective lightings are displayed in Figure 2.22.

In making Figure 2.22, we repeated the procedure six times. The pictures made in each repetition are grouped together to give a sense of the fractal pieces from which the tree is composed. Each piece is a copy of the whole tree but shrunken (and distorted), possibly rotated, and possibly reflected. Note that the trunk is made of two copies of the tree—both shrunken by a lot in the horizontal direction, one standing upright, the other reflected across the x-axis. Look at the parameter table. Can you see how rules 1 and 2 do exactly what is displayed in the figure? The four branches are much clearer copies of the whole tree,

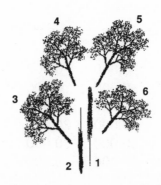

Figure 2.22   *The tree shown in Figure 2.21 pulled apart.*

though each has suffered some distortion through unequal scalings (unequal R and S).

In addition to "tree," two other plant images commonly generated by IFS are "fern" and "lace" (as in Queen Anne's). Of course, none of these forgeries is quite right; each is a little too perfect, a little too symmetrical. Nonetheless, they seem to capture the essences of "treeness," "fernness," and "laceness" remarkably well. Tinkering with the parameters some and possibly adding a few more transformations to the list would probably lead to even more realistic images.

Six not-so-obvious rules enable us to make a decent picture of a tree. This situation may be fairly general. Producing natural fractal images may require only a relatively few transformation rules, but what they are will probably not be so transparent. And, as we have commented before, they are not likely to be unique.

Here, then, is an interesting question: Is there an efficient and reliable algorithm for generating an optimal set of transformations that can be used to produce a suitably accurate image (or composite of images) of a given natural fractal (or collection of natural fractals)? Let's put this slightly differently. Suppose we have a photograph of a landscape, or a family picnic, or even some text. Are there ways of readily chopping the photograph into fractal pieces and encoding optimal transformation rules for each piece so that at some later time it can be reproduced by computation? (That is, can we economically store the photograph in the computer?)

A partial answer to this question is supplied by the so-called *Collage Theorem* [70]. The Collage Theorem is a kind of generalization of our discussion on page 49 about generating images of self-similar mathematical fractals. In that discussion, we said that to produce a picture of a self-similar fractal, we must identify within the fractal N smaller copies of the whole thing. The transformation rules needed to create those N copies will, when iteratively applied (deterministically or randomly), suffice to generate the desired picture. The Collage Theorem takes a broader view. It says, start with *any* picture — not necessarily a symmetrical mathematical fractal. Using only smaller copies of the original — perhaps rotated, translated, and squashed by affine transformations — try to cover the original with a collage of these copies without too much "slop" or too many holes. (Of course, the more little copies you use, the better covering you will be able to make if the original is not exactly symmetrical, as is the case for pictures of natural objects.) The copies used in the covering define an IFS scheme for generating an approximate image of the original picture by computer. That is, record R, S, Theta, Phi, E, and F for each copy used in the covering. Run this set of parameters in an iterated function systems program, and an approximate replica of your starting picture will emerge! (The better the covering, the better the replica.)

Let's try it. The left-most picture in Figure 2.23 shows a scanned image of a real oak leaf. In the middle picture we have covered the original with five smaller copies of itself, each mangled by some affine transformation. (The "stem," the small piece 5, is a little difficult to see.) The parameter values of the transformations were then used in an IFS program to generate the right-most picture. The parameters corresponding to the covering copies are recorded in the accompanying table.

The results of our little experiment are very interesting. First, note that the covering we chose was pretty crude; some of the underlying leaf peeks through the covering on the edges, and some in the interior. Nonetheless, the IFS output is not just an unidentifiable blob. It clearly mirrors the shape of the original leaf. Because the covering contains some holes, the IFS picture

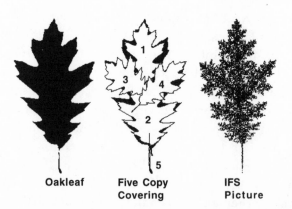

Oakleaf   Five Copy   IFS
                 Covering   Picture

Figure 2.23   *An oak leaf, a covering, and an IFS reconstruction.*

No.	R	S	Theta	Phi	E	F	Prob
1	0.500	0.500	0.000	0.000	−0.051	1.000	0.290
2	0.750	0.500	0.000	0.000	0.000	0.092	0.340
3	0.500	0.400	53.000	53.000	−0.059	0.924	0.180
4	0.400	0.500	−50.000	−50.000	−0.168	0.723	0.180
5	0.015	−0.250	0.000	0.000	−0.008	0.202	0.010

does too. And because the IFS picture is a fractal object, the holes repeat everywhere, on different length scales. The effect is to produce an oak leaf that looks like it's been attacked by Japanese beetles. Still, the close proximity of this quickly produced image to a thing of nature is undeniable.

IFS codes for making natural fractal images may be succinct, but they aren't trivial. Complicated images usually aren't well described by a single fractal. Typically, an image has to be broken into pieces and an IFS code found for each piece. After you've played with trying to make good IFS images of natural objects for a while, you will be impressed at how difficult it is to generate really accurate and at the same time efficient, transformation rules. Michael Barnsley seems to have found a very efficient algorithm for using the Collage Theorem for this purpose, though what it is in detail remains a closely guarded secret. Barnsley has established a private company (Iterated Systems, Inc.) to develop products exploiting his algorithm. A number of these are already on the market.

Because of the huge data compression associated with IFS image generation, the communications industry is extremely interested in these techniques. A rather crass, but nonetheless potentially profitable, application might be high-definition television (HDTV), which, at the present state of development, crams about twice as much information on a typical screen as does regular TV. ("Information" doesn't have to be useful.) One can imagine the transmitting of ordinary video pictures along with IFS codes, both of which would be received by the TV. A small dedicated IFS processor might be able to flesh out the ordinary signals by processing the IFS code in real time.

Coloring

Practical utilization of IFS techniques requires not only getting
the outline of objects right, but also somehow encoding the
proper coloring of the generated images. Here is a very simple
way of coloring IFS generated pictures. To help visualize the
procedure, we start with a fractal in which the pieces are well
separated from each other. Let's take the triangular dust depicted in Figure 2.6 as a specific example. Remember, that
picture was associated with three transformations:

1 = Scale by 0.333.

2 = Scale by 0.333; translate +0.5 in x.

3 = Scale by 0.333; translate +0.5 in y.

Figure 2.24 shows (a) the triangular dust and (b) a second copy
with pieces surrounded by circles of different sizes. If we take
the entire dust in part a and transform it by **1** (respectively, **2**, **3**),
then we get the shrunken dust contained within the circle labeled **1** (respectively, **2**, **3**) in part b. That is, we can think of the
dust within the circle labeled **1** as being the "**1** part of the whole
picture." If we apply **3** to the whole picture a, and then apply **3**

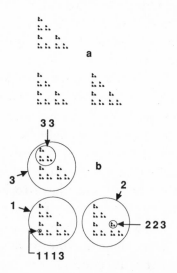

Figure 2.24   *A triangular dust (a) and the labeling of some of its parts (b).*

again to what results, we get the even more shrunken dust contained within the circle labeled **33**. That small copy can be thought of as the "**3** part of the **3** part of the whole picture." Similarly, applying the sequence **3**, then **2**, then **2** produces the dust within **223**; that is, it's within "the **3** part of the **2** part of the **2** part." The label **223** tells us where to look: first in the largest **2** piece, then in the **2** piece of that, and then in the **3** piece of that. Do you see how the label **1113** comes about?

Now think of creating the dust by using the random IFS method (starting somewhere on the fractal). Every point within the circle labeled **1** came about because the transformation **1** was the last transformation selected. Similarly, all points within **33** came from points whose last two transformations were **3** and, immediately before that, **3**. All points within **223** came as the result of **2** being the last transformation applied, preceded immediately by **2**, preceded immediately before that by **3**.

Every point in our fractal picture has a unique address. The address starts with the biggest subpiece, then the next biggest, then the next, and so on. But this address is also the sequence of transformations that led to that point being placed there as a result of playing the random IFS game. Thus, if we wanted to color all the points within the large circle **1** red, say, all we would have to do is assign red to any pixel lighted as a consequence of **1** being the last transformation applied. To color all the points within **223** green, we would have to save a record of the last three transformations; any time **2**, the last transformation applied, followed **2** that itself followed **3**, that pixel would be colored green.

Obviously, this coloring trick can be applied to any IFS construction. The location—the address—of every point in the construction is the sequence of transformations (in reverse order: last first, first last) necessary to produce that point. Coarse colorings can be achieved by keeping track of only the last few transformations. Coloring of finer pieces requires remembering longer strings. This technique was used to generate the Cherry Tree of Plate 13, which is the familiar "tree" IFS with some coarse and some fine pieces lit with different colors.

Another method of coloring involves adjusting the probabilities. In Figure 2.20, we saw that different probability assignments can affect the rate at which regions of the picture are filled in by the random IFS algorithm. Colors can be assigned to a region by counting how often that region is visited by the randomly dancing IFS point, and this is governed by the probabilities assigned to the transformations.

What Is Random?

The word *random* has appeared without definition several times in the first two chapters. You probably have an intuitive feeling that random implies unpredictability. It does, but that's not enough; many processes are unpredictable, yet we wouldn't call them random. For example, the first time we see a movie, we cannot predict what the actors will say next (unless the movie is really awful). By the hundredth time, though, the dialogue contains no surprises at all. What the actors say is determined by rules — a script — that we eventually learn. Of course, some scripts are more difficult to figure out than others. A script in a language we don't speak takes a lot longer to understand. Nonetheless, it is possible, in principle at least, to predict what will occur when there are underlying rules that determine the action. For our purposes, we take a narrow definition of *random:* unpredictable *even in principle*. In the absence of underlying rules, the most compact description of a random sequence of events is the sequence itself; no shorter description is possible.

Note, though, that random does not mean "anything goes." A random coin toss produces a head or a tail only. A random number sequence might well be a string of fractional values between 0 and 1. Random outcomes usually are bounded in some way. Also, a random sequence does not have to be uniform. *Uniform* means that a long enough random sequence will have equal numbers of each possible outcome, equal numbers of all possible successive pairs of outcomes, equal numbers of all possible successive triples of outcomes, and so on. A nonuniform random sequence contains all possible outcomes, all possible pairs, all possible triples, . . ., but the outcomes and the pairs and the triples and . . . don't have to appear equally often. A die with a "head" painted on five sides and a "tail" on the other can still produce a random sequence of heads and tails, but in that sequence, heads will appear more often than tails; head–head, head–tail, and tail–head pairs more often than tail–tail pairs; and so on.

When we create an IFS picture with a uniform random sequence of rule selections, every portion of the picture gets colored in, including those very fine parts that require long strings of all the same rule (such as near the vertices in the Gasket). When we use a nonuniform random sequence, some parts are filled in faster than others—that's what we saw in Figure 2.20—but eventually all points on the picture get hit.

It's interesting to ask what happens when we use a sequence of rule selections that is not random. What happens to the IFS picture then? As an example, let's try the rules we used before for making a square:

1 = Scale by 0.5.

2 = Scale by 0.5; then translate by +0.5 in the x-direction.

3 = Scale by 0.5; then translate by +0.5 in the x-direction and by +0.5 in the y-direction.

4 = Scale by 0.5; then translate by +0.5 in the y-direction.

And let's select the rules in the order **123412341234 1234**. . . . When these four rules are selected uniformly randomly, the IFS picture constructed is a uniformly filled-in square. Clearly, in the string shown each rule will be selected equally often, but because not all pairs, not all triples, . . ., appear in this string, the whole square will not fill in. To see what does happen, we can use the idea of addresses. See Figure 2.25. We have chopped up a square into relevant parts. Look at the **4** part of the square. The rule **4** is always preceded by **3**, so the IFS eventually will have points only inside the **3** part of the **4** part (labeled **43** in the figure). But **2** always precedes **3**, so points will only fall within the **2** part of the **3** part of the **4** part (labeled **432**). But **1** always precedes **2**, so. . . . We soon see that only one point will be hit in the **4** part of the square—within the "dot" shown there. Similarly, only one point will be hit in each of the three other main parts of the square—again as shown. *The*

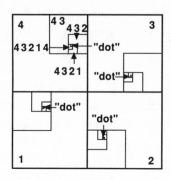

Figure 2.25 *The IFS picture constructed by the cycle* **12341234**
12341234 . . . .

IFS picture is just four isolated points. Note that the points indicated
are the ones *ultimately* hit. If you don't start with a point on the
ultimate picture, it takes a bit of time to contract from where
where you start down to the points indicated. The sequence
123412341234 . . . is a periodic cycle that repeats exactly
after four steps: a so-called 4-cycle. Extrapolating from this
example, we state the plausible result that cycles of length N
produce IFS pictures consisting of N isolated points. Of course,
the sequence **123412341234** . . . repeats exactly after 8,
12, 16, . . . steps. Thus what we mean is that a cycle of shortest length N (4, in this example) produces an IFS picture consisting of N isolated points.

Incidentally, the idea of addresses enables us to see more
clearly why the random and the deterministic IFS algorithms
produce the same picture. Single pixels in the picture are just
little squares with finite addresses. Suppose, for example, the
"dot" in the upper left portion of Figure 2.25 is on the IFS
picture made by the deterministic algorithm. That dot then has
the address **4321432**. (Can you see why?) How do we know that
if the picture is made by the random algorithm, that pixel will
actually be lit? Start from any other pixel on the picture and
apply rule **2** to that pixel. The new pixel lit (also on the picture)
then has an address starting with a **2**. Suppose rule **3** is applied
next. The next pixel now has an address starting with **32**.
Similarly, if the rules **4**, **1**, **2**, **3**, and **4** come up, in that order,
the last pixel lit has an address starting with **4321432**; that is, it

is in the desired "dot." Because the random algorithm generates all strings of all lengths of the integers **1, 2, 3, 4**, the string **2, 3, 4, 1, 2, 3, 4** will come up sometime. In fact, if we say a pixel on the picture is a square with a seven-digit address, the random algorithm will hit all seven-digit strings—and hence will light all pixels on the picture—if we wait long enough.

One can imagine sequences that are intermediate in complexity between uniform randomness and exact periodicity. For example, suppose we choose the four square rules uniformly randomly but, in addition, require that whenever **3** occurs we will not allow **1** to occur next. Then the picture that forms will have an empty square part corresponding to the **3** part of **1**. Of course, the **3** part of the **1** part of every part of the square will also be empty, creating a kind of Swiss cheese effect. A few of these empty squares are shown in Figure 2.26.

In Chapter One we showed what happens when we play the Chaos Game (Four Corners, One-half Version) using the sequence of bases along a DNA strand. The Chaos Game is a random IFS construction, so the structure we observed in the DNA Chaos Game should be analyzable by addresses. Figure 2.27 redraws Figure 1.23, the Chaos Game result for the enzyme amylase, but with subsquares marked off that are almost totally free of any fill. These subsquares are in precisely the same positions as in Figure 2.26. The inference is that the DNA strand has almost no pairs (the marked windows are not completely empty) in which

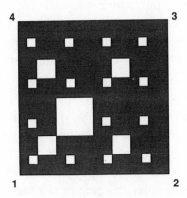

Figure 2.26   *Some holes in an otherwise filled-in square made by the random IFS algorithm plus the condition that rule* **1** *cannot follow rule* **3**.

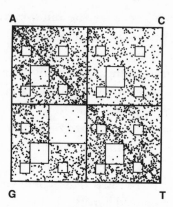

Figure 2.27   *Amylase DNA IFS square with empty regions marked.*

the base G follows the base C. Furthermore, it is easy to see a dark diagonal connecting the A corner and the T corner. This diagonal suggests that in the strand there are lots of A's and lots of T's (because the vertices are well filled in) and lots of combinations of A's and T's in groups. What other information might be read off the IFS construction is not immediately obvious, but it is clear that IFS pictures can be used to determine quickly whether, in some cases at least, the sequences used to construct them result from processes with hidden rules.

When the random IFS algorithm is used to make pictures, the "random number generator" of the computer is used to generate rule selection sequences. The random number generator is designed to produce approximately uniform random sequences. The computer doesn't actually have access to real random numbers. Instead, it employs a process whose underlying rules are so complex that prediction is extraordinarily difficult—though in principle still possible. Usually the generation process requires a starting, or "seed," value. When the same seed value is used, the same output results. The process is completely deterministic. Such number sequences are said to be pseudo-random. How well they approximate uniform randomness can be tested by the random IFS algorithm. If strings so generated color a square uniformly, we can say that the sequences are *as good as* uniformly random. In this sense, IFS construction can provide an operational test of functional randomness. We'll come back to the very interesting application of IFS picture-making techniques as a test of randomness later, when we study noise.

Speculation

We close this chapter with a couple of observations and related speculations. We note first that, as the name implies, iteration is at the heart of how the iterated function systems method works. That is, IFS generates new pictures from old by applying the same set of transformations repeatedly. In terms of the real-time emergence of an IFS-generated picture on a computer display, we can say that the current state of the display screen (what pixels are currently lit) evolves from the immediately previous state by the application of the transformation rules that define the IFS. In turn, the next screen is determined by the current screen through another application of the transformation rules.

We note further that the IFS construction of images cannot be simply inverted. Any *one* rule in the set can be exactly inverted; the inverse rule is (a) subtract the translations, (b) rotate by the same angles in the opposite directions, and (c) stretch by the reciprocals of the scaling factors. If there are N rules in the set, however, simultaneous application of the inverses to every point on a current picture — as is required by the deterministic algorithm — produces N new points, only one of which corresponds to a point on the actual preceding image. The situation is no better in the random algorithm; there, one of the N rules is selected at random in each screen update. Thus only one in N times will this process hit the correct previous point if we use the N inverse rules; $N - 1$ times, spurious points will be generated. The noninvertible character of IFS indicates that the process is nonlinear. Figure 2.28 shows what we mean.

The lefthand graph is supposed to represent a situation in which the "present" of some system (such as the computer display screen) is related in a linear fashion (represented by the bold line) to the past state. In that case, the state that gave rise to the new state is uniquely determined. The righthand graph depicts a nonlinear relationship (represented by the curve) between present and past states. In this picture, many present states could

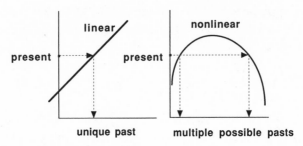

Figure 2.28   *Predicting the past.*

have arisen from either of *two* past states. Even though in each example the present is uniquely determined by the past, the inverse is not true. A nonlinear relationship between past and present cannot be uniquely inverted; in that case, the past that led to the present is ambiguous.

Well, these observations prompt the following, as yet incompletely answered, question. Extraordinarily natural-looking images can be generated via IFS techniques — that is, by iterated application of simple (though nonlinear) rules. Do natural objects similarly form as the result of relatively simple (though perhaps nonlinear) iterated processes? Some titillating evidence suggests that the answer to this question may well be yes. In this event, we would expect that the most succinct and accurate description of complicated natural objects — mountains, river tributary networks, clouds, plants, circulatory systems, neural connections, geological resource distributions, and so on — would be in terms of a few suitably chosen rules rather than in terms of complicated geometrical structures (that is the usual way of attempting to characterize such objects). Imagine the data compression involved in such "evolutionary" descriptions! One is even led to fantasize that at least some aspect of genetic coding might be IFS-like. It is now known that DNA isn't just a linear code — a string that defines all the organism's proteins and dictates where to put them. Parts of the DNA molecule turn other parts on and off in some complex biochemical interplay that is only vaguely understood. One wonders if somewhere within this complex of internal interactions, DNA contains some IFS-like meta-code, consisting of some iteratively applied assembly rules, that serves to make the wonderfully complex organic machines of life. The human genome is likely to be mapped out in detail in the near future, so we will probably know soon whether this speculation has merit.

c h a p t e r

3

Unearthly Earthly Dimensions

THE SPIKY, RED pinnacles of Bryce Canyon in southern Utah have a self-similar quality. So do the gently humped Blue Ridge Mountains of central Virginia. The pinnacles and the mountains are natural fractals. Despite sharing the property of self-similarity, however, these geological structures would never be confused for each other, even if they were made to have the same color. How can we describe what makes these fractals different? In this chapter, we discuss one attempt to measure the difference between fractals: through their dimensions. As we will argue, fractal dimensions can have important consequences. It may even be that we would not be here—no people, no animals, no plants, no planet Earth!—were it not for the fractal dimension of dustballs in the earliest days of our solar system.

Simplicity and Complexity in Classical Geometry

The objects of classical geometry—straight lines, circles, spheres, cones, and so on—are "simple." That is, each can be represented (modeled) by simple equations that describe the shape and extent of the object. Let's take as an example a straight line. One way of modeling a straight line is to embed the line in a coordinate system, as in Figure 3.1. An important characteristic of a straight line is that it has a constant slope. If we take any two

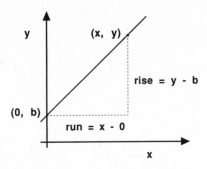

Figure 3.1   *A straight line.*

points on the line, then the slope is the "rise" of the line between the chosen points (the difference in their y-values) divided by the "run" associated with that rise (the difference in the points' x-values). In other words, if we choose as our points the points (0, b), where the line goes through the y-axis, and any other point (x, y), we can represent the slope, m, by the ratio

$$m = \frac{y - b}{x - 0}$$

Rearranging these terms yields

$$y = m \cdot x + b$$

which is the familiar equation for a straight line. Other simple curves "living" in the x–y plane—such as circles, ellipses, parabolas, hyperbolas, and so on—similarly can be described by their own equations relating the x- and y-values of points on the curves. For example, the coordinates of points on a circle whose center is at $x = 0$, $y = 0$, and whose radius is equal to 1, are related by

$$x^2 + y^2 = 1$$

Now, imagine a geometrical object living in the plane that is slightly more complicated, the "tent" you see in Figure 3.2. At first it may not be obvious that this object is more complicated.

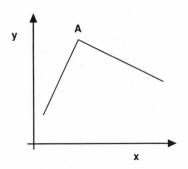

Figure 3.2   *Two lines making a tent.*

But think for a moment how you might model it. One way of doing so is to break the tent up into pieces: to the left of A there is one straight line described by its characteristic equation, and to the right is a second described by *its* equation. Thus we would have the following rather homely, but nonetheless correct, description:

$$y = m_{left} \cdot x + b_{left} \qquad \text{if x is "to the left of A"}$$
$$\text{and} \quad y = m_{right} \cdot x + b_{right} \qquad \text{if x is "to the right of A"}$$

A composite object consisting of two line segments requires two equations to describe it. If we drew three line segments, making something akin to an N, then we would need three equations patched together, and so on. Note that these objects all have sharp kinks in them; the presence of kinks is precisely what makes them more complicated than the simple straight line.

A similar situation would arise if we considered objects with breaks in them like the one in Figure 3.3. In this figure two line segments are displayed, both with the same slope; the termination point, A, of the first has the same x-value as the beginning point, B, of the second. Just as in Figure 3.2, two equations are required to describe this object.

The simple curves of classical geometry are said to be smooth. We say that a curve is *smooth* if, when we magnify the curve sufficiently at any point, it looks like a little piece of a straight line. In this sense, a smooth curve is locally indistinguishable from a straight line. (Similarly, we say that a surface is smooth if

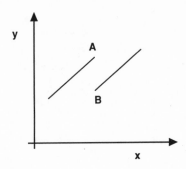

Figure 3.3 *A broken line.*

any point on it looks like a little flat patch of a plane.) The objects shown in Figures 3.2 and 3.3 are not smooth. At their "break" points, they are both *two* straight lines.

An alternative way of saying that a curve is smooth is to say that at any point, such a curve has a unique tangent line. The *tangent line* to a curve at a point nicks the curve at that point. If we magnify the curve at the point where it's being nicked sufficiently, we can't tell the difference between the curve and its tangent line. This situation is shown in Figure 3.4. Smooth curves "inherit" global analytic descriptions from their local linearity; they can be described by simple equations.

Note that the object in Figure 3.2 fails to have a unique tangent line at the point A because assignment of the tangent line there depends on how the point is approached. If we approached A along the line on the left, we would say that the tangent line at A has the slope m_{left}, whereas approaching from the right would suggest a tangent line with the slope m_{right}. The

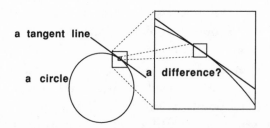

Figure 3.4 *The tangent line to a point on a circle at different magnifications.*

object does not look like a single straight line at the kink point. A sharp kink point can have no unique tangent line attached to it. A similar problem occurs on the object in Figure 3.3 at the break point. Assignment of the tangent line depends on how the break is approached; at the break point the object does not at all look like a single straight line. The break is a point of *discontinuity;* there can be no unique tangent lines at discontinuities.

It should now be clear why objects such as the Koch Curve were considered "monsters" by classical mathematicians. The Koch Curve consists only of kink points. It has no tangent line anywhere. It is totally unsmooth. Magnifying it does not get rid of its wriggliness. It continues to wriggle, sharply, on all levels of magnification; nowhere is it straight (see Figure 3.5).

The Cantor Set is also totally unsmooth because it is completely discontinuous, being a dust of disconnected points. Though at some level of resolution the crowded points near the ends of the Cantor Set seem to make little line segments, increasing magnification reveals that these little lines continually fall apart into finer and finer disconnected pieces (see Figure 3.6).

In general, the "fractured" character of fractals—be they highly self-similar, like the Koch Curve or the Cantor MTS, or only statistically self-similar, like those found in nature—leads to such jagged and broken structures that they are essentially nowhere smooth. And because of their complete lack of smoothness, the geometrical structures of fractals cannot be described by simple equations. Of course, we can often supply simple iterative rules for manufacturing such objects, but we cannot usually answer the elementary question "Does the point (x, y) lie on such-and-such fractal?" This question is a "proper" one for classical geometrical objects, but it is totally inappropriate for fractal monsters.

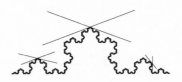

Figure 3.5   *The Koch Curve has no tangent line anywhere.*

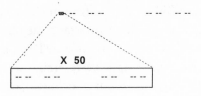

Figure 3.6   *End piece of the Cantor Middle Thirds Set magnified by a factor of 50.*

A Dimension Primer

If a defining equation cannot be found for fractal objects, is there *anything* that can be borrowed from classical geometry and used to characterize these beasts? Fortunately, the answer to this question is yes. One such property is the dimension of the fractal.

What do we mean by dimension? Figure 3.7 reviews some primitive notions concerning dimension with which you are probably familiar. The objects shown in Figure 3.7 are from classical geometry. They are examples of *Euclidean objects,* locally smooth except at their edges and corners. Geometrical objects that are locally pointlike (that is, they look like isolated points when magnified enough) have dimension 0, just like a single point. Objects that are locally like a line (smooth curves) have dimension 1, just like a line. Objects that are locally like the insides of little square patches (smooth surfaces) have dimension 2, just like a square. Objects that are locally like the insides of a cube (smooth volumes) have dimension 3, just like a cube.

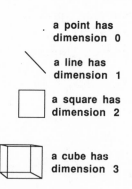

Figure 3.7   *A few elementary dimensions.*

How do we know lines have dimension 1 and squares dimension 2? Here's a physically useful way to think about this question. Let's imagine we have a physical model of a line — a stiff, uniformly made wire whose thickness is ignorable. We require that it be uniformly made because we want each part of the line to be identical to every other part. For the sake of numerical simplicity, let's declare that the length of the wire is 1 "unit of length." This wire weighs something; let's similarly declare its weight (or more properly its mass) to be 1 "unit of mass." Using these units relieves us from having to worry about real units such as meters and feet and pounds and grams. If we cut our unit wire into two equally long, smaller pieces, each smaller piece has a length of $\frac{1}{2}$ unit and a mass of $\frac{1}{2}$ unit. Cut the wire in half again, and each new piece is $\frac{1}{4}$ unit long and has a mass of $\frac{1}{4}$ unit. There's a systematic relationship between the mass, M, of a piece of wire made identically to the original and its length, L. It is

$$M = L$$

as long as we measure mass and length in the units defined.

Suppose now that we start with a uniformly made square sheet. Whereas the wire had a unique length, a sheet does not. As long as we use the analogous length in each copy, it doesn't matter what length we use, however. So let's make life easy for ourselves and use the length of a side. We declare that a sheet of side length 1 has a mass of 1. Cut the sheet into pieces, each of length equal to $\frac{1}{2}$. Because there are 4 smaller squares within the first, each $\frac{1}{2}$ unit of length on a side, the mass of one of these smaller squares is $\frac{1}{4}$ (see Figure 3.8). Similarly, a square that is $\frac{1}{2}$

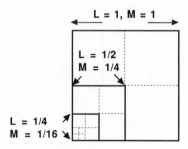

Figure 3.8 *Successive squares each $\frac{1}{2}$ as long as the previous one.*

unit long has four squares of length $\frac{1}{4}$ in it, so the mass of a square of length $\frac{1}{4}$ is $\frac{1}{16}$ (that is, $\frac{1}{4}$ of $\frac{1}{4}$). Each time we reduce the length by a factor of $\frac{1}{2}$, we reduce the mass by a factor of $\frac{1}{4}$. We infer that the relationship between mass and length this time is

$$M = L^2$$

because $1 = 1^2$, $\frac{1}{4} = (\frac{1}{2})^2$, $\frac{1}{16} = (\frac{1}{4})^2$, and so on.

Finally, let's look at a uniform cube, side length equal to 1, mass equal to 1. This cube consists of 8 cubes of side length $\frac{1}{2}$, so an identically made cube with length equal to $\frac{1}{2}$ has a mass equal to $\frac{1}{8}$ (see Figure 3.9). Similarly, if we cut out of this smaller cube an even smaller cube of length $\frac{1}{4}$, *its* mass will be $\frac{1}{8}$ of $\frac{1}{8}$, or $\frac{1}{64}$. Each time we reduce the length by a factor of $\frac{1}{2}$, we reduce the mass by a factor of $\frac{1}{8}$. The mass–length relationship for such cubes is

$$M = L^3$$

because $1 = 1^3$, $\frac{1}{8} = (\frac{1}{2})^3$, $\frac{1}{64} = (\frac{1}{4})^3$, and so on.

Note that the relationship for the line, $M = L$, can also be written as $M = L^1$. We can summarize the three cases examined above by stating this general rule: the *dimension* of a Euclidean object is the power to which its length has to be raised in order to get its mass (if mass and length are measured in convenient, but unconventional, units). In other words, the physical process of relating mass and length enables us to operationally measure

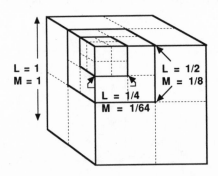

Figure 3.9   *Successive cubes each $\frac{1}{2}$ as long as the previous one.*

dimension. This whole process works equally well for smooth curves (dimension 1), smooth surfaces (dimension 2), and smooth volumes (dimension 3) of all shapes, provided we can make a series of identical copies of different sizes.

Next we turn our attention to a physical model of a Gasket. Imagine that the model is made of thin, stiff wires. We want to measure off some lengths and associated masses in order to deduce the Gasket's dimension. As with the lines, squares, and cubes, start with a unit Gasket, length = 1, mass = 1. (Choose a convenient length, such as the length of the Gasket's height.) In such a Gasket there are three identical copies with length = $\frac{1}{2}$. Each of these copies therefore has mass = $\frac{1}{3}$ (see Figure 3.10). Identical copies $\frac{1}{4}$ in length have mass $\frac{1}{9}$; those with length $\frac{1}{8}$ have mass $\frac{1}{27}$. Every time we reduce the length by a factor of $\frac{1}{2}$, we reduce the mass by a factor of $\frac{1}{3}$. We seek a mass–length relationship of the same form as those we noted above:

$$M = L^d$$

where d is the "dimension" of the Gasket. Well, trivially $1 = 1^d$; that doesn't tell us much about d. But d has to be just right so that $\frac{1}{3} = (\frac{1}{2})^d$, $\frac{1}{9} = (\frac{1}{4})^d$, $\frac{1}{27} = (\frac{1}{8})^3$, and so on. Clearly, d is going to be a little strange. For example, raising $\frac{1}{2}$ to the 1 power gives $\frac{1}{2}$, which is bigger than $\frac{1}{3}$. On the other hand, $(\frac{1}{2})^2 = \frac{1}{4}$, which is smaller than $\frac{1}{3}$. Thus the proper value of d has to be between 1 and 2!

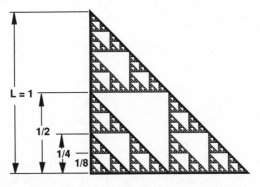

Figure 3.10   *Successive Gaskets each $\frac{1}{2}$ as long as the previous one.*

The dimension d, measured as we have done, is an exponent. To solve for d — that is, to isolate d by itself — we have to "bring it down from the exponent." This is exactly what the logarithm function enables us to do. Taking the logarithm of both sides of the express $\frac{1}{3} = (\frac{1}{2})^d$ leads us to the rather amazing result

$$d_{Gasket} = \log(3)/\log(2) = 1.58496 \ldots$$

(which, as we surmised, is indeed between 1 and 2).

The Logarithm Function

The logarithm function assigns a unique output number to any input number. Formally, this assignment is represented by

$$\log(x) = y$$

This relationship — which is read "log of x equals y" — assigns the output y to the input x through the logarithm recipe. It has two essential properties:

$$\log(x \cdot y) = \log(x) + \log(y) \qquad (1)$$
$$\log(x^p) = p \cdot \log(x) \qquad (2)$$

From these two properties, we can derive the useful consequences

$$\log(x/y) = \log(x \cdot [1/y]) = \log(x \cdot y^{-1}) = \log(x) - \log(y)$$

and

$$\log(1) = \log(x/x) = \log(x) - \log(x) = 0$$

The actual value of the logarithm of a specific number depends on the base of the logarithm. The base of a logarithm is that value of x for which $\log(x) = 1$. If it is absolutely necessary to know what base the logarithm is referred to (often it is not), then we compulsively write $\log^x(\ )$ which means "logarithm in the base x." Logarithms that are referred to different bases differ from each other only by numerical multiplicative constants.

Here's an example: How are $\log_{10}(\ )$ and $\log_2(\ )$ related? The solution: Any number y can be written as 2^q.

$$y = 2^q$$

Now, according to equation (2),

$$q = \log_2(y)$$

But taking $\log_{10}$ of both sides of the expression for y yields

$$\log_{10}(y) = q\ \log_{10}(2)$$

so

$$\log_{10}(y) = \log_2(y)\ \log_{10}(2)$$

Thus log base 10 of anything is log base 2 of that thing times the numerical constant $\log_{10}(2)$. Because powers of 10 are so important to us, $\log_{10}(\ )$ values are well tabulated, and hand calculators often have "log" buttons that really mean $\log_{10}$. The number $\log_{10}(2)$ is readily found; on a calculator you enter "2" "log." The answer is about 0.30103 . . . .

Because logarithms in different bases differ only by multiplicative constants,

$$\frac{\log_{[\text{any base}]}(x)}{\log_{[\text{any base}]}(y)} = \frac{\log_{10}(x)}{\log_{10}(y)}$$

and because ratios of logs are what we need for dimension calculations, we can delete reference to the base without ambiguity. We can use base 10 for convenience in calculating actual values.

How do we get $d_{\text{Gasket}} = \log(3)/\log(2)$? Start with $\frac{1}{3} = (\frac{1}{2})^d$. Taking the log of both sides yields $\log(\frac{1}{3}) = d \cdot \log(\frac{1}{2})$, but $\log(\frac{1}{3}) =$

$-\log(3)$ and $\log(\frac{1}{2}) = -\log(2)$, so $d = [-\log(3)]/[-\log(2)] = \log(3)/\log(2)$. The exponent is supposed to ensure that $\frac{1}{3} = (\frac{1}{2})^d$, as well as $\frac{1}{9} = (\frac{1}{4})^d$, $\frac{1}{27} = (\frac{1}{8})^d$, and so on. Does it? Note that each successive mass–length relationship can be written

$$(\tfrac{1}{3})^n = [(\tfrac{1}{2})^n]^d$$

where n is 1, 2, 3,. . . . . Taking logs produces

$$d = \frac{-n\log(3)}{-n\log(2)} = \frac{\log(3)}{\log(2)}$$

without regard to which entry in our list of masses and lengths we use.

Some Meaning

Straight lines, squares, cubes, and Gaskets can all be thought of as being made of exact, smaller copies of themselves, nonoverlapping except at corners or at edges. (In this sense, straight lines, squares, and cubes — Euclidean objects all — are perfectly good fractals as well.) A straight line is made, for example, of two straight lines, each shrunk by a factor of $\frac{1}{2}$. A square is made of four squares, each shrunk by a factor of $\frac{1}{2}$. A cube is made of eight cubes, each shrunk by a factor of $\frac{1}{2}$. A Gasket is made of three Gaskets, each shrunk by a factor of $\frac{1}{2}$.

The dimension of an object made of N exact (distortion-free), nonoverlapping, smaller copies of itself, each shrunk by a factor of s, can be written down immediately. It is

$$d = \frac{\log(N)}{\log(1/s)} \tag{3}$$

For a line,

$$d = \log(2)/\log[1/(\tfrac{1}{2})] = \log(2)/\log(2) = 1$$

For a square,

$$d = \log(4)/\log[1/(\tfrac{1}{2})] = \log(4)/\log(2) = \log(2^2)/\log(2) = 2\log(2)/\log(2) = 2$$

For a cube,

$$d = \cdots = \log(8)/\log(2) = \log(2^3)/\log(2) = 3$$

Finally, for a Gasket, $d = \cdots = \log(3)/\log(2)$

We know of some other fractals of the kind for which this simple recipe works. For example, the Koch Curve (Figure 2.10) is composed of four identical pieces, each shrunken from the original by a factor of $\tfrac{1}{3}$. Consequently,

$$d_{\text{Koch Curve}} = \frac{\log(4)}{\log(3)} = 1.26185. \ldots$$

The Cantor Middle Thirds Set (Figure 2.7) is two pieces, each shrunken by a factor of $\tfrac{1}{3}$ from the original. Thus,

$$d_{\text{Cantor MTS}} = \frac{\log(2)}{\log(3)} = 0.63092. \ldots$$

The nice integer dimensions of Euclidean objects tell us something about those objects. The dimension is related to the "proper measure" of the object. A line is properly characterized by a one-dimensional measure, its length; a square by a two-dimensional measure, a [length × length]—that is, by an area; a cube by a three-dimensional measure, a [length × length × length]—that is, by a volume. Any one-dimensional object, however winding and twisting, can be imagined to be broken up into a collection of (perhaps small) straight lines, and the length of that object is just the sum of all the lengths of the segments of which it is made. Any two-dimensional object, however filled with holes, can be thought of as a patchwork of squares—some perhaps very tiny, filling in nooks and crannies—and its area is just the sum of the areas of the squares in the patchwork. Any three-dimensional object similarly can be imagined as filled with cubes of various sizes, and its volume is the sum of the volumes of those cubes.

What happens if we "mismeasure" a Euclidean object? Suppose, for example, we ask for the "total length" of a square. What would that be? To measure the length of a curve, we completely cover the curve with little straight lines, trying not to leave gaps or to do much overlapping: we want to approximate the curve as best we can by a sequence of lines. When we have a good covering, we just add up the lengths of all of the lines needed. That's what we would have to do to measure the total length of a square. See Figure 3.11. The covering starts with one line of length 1 stretched across the square's midsection. Of course, this is a pretty poor covering because there is a lot of square that is not covered. We need more lines. Let's add them systematically. Imagine the square divided into 4 subsquares each with side length equal to $\frac{1}{2}$. Place a line across the midsection of each of the subsquares and add up the total length of these 4 segments: $4 \times (\frac{1}{2}) = 2$. There's still a lot of exposed square, though, so imagine the original square divided into 16 subsquares, each of side length equal to $\frac{1}{4}$. The total length of the analogous covering is $16 \times (\frac{1}{4}) = 4$. The next finer covering has a total length of $64 \times (\frac{1}{8}) = 8$. And so on. The finer the covering, the more total length there is, and worse, the total length doubles each time. In the limit where we have lines as close as we like to any point in the square, with no holes showing, the total length is infinite.

On the other hand suppose we ask, "What is the volume of a square?" To measure volume properly, we need to cover the square with little volumes with no holes and no overlaps (see

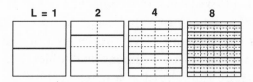

Figure 3.11   *Covering the square with straight line segments.*

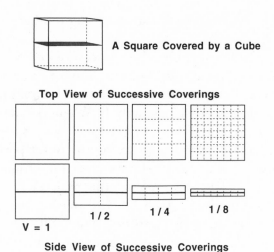

A Square Covered by a Cube

Top View of Successive Coverings

Side View of Successive Coverings

Figure 3.12 *Covering the square with cubes.*

Figure 3.12). To cover a square with cubes, we insert the square into the midsections of the cubes. One cube covering the square has a volume equal to 1. But such a covering has too much slop: too much cube above and below the square. (See the side view of the leftmost covering in the figure.) We need to tighten up the covering, so we try 4 cubes each $\frac{1}{2}$ on a side (the next covering from the left in the figure). This covering yields a total volume of $4 \times (\frac{1}{2})^3 = \frac{1}{2}$. This is better, but there is still too much slop. With 16 cubes of side length $\frac{1}{4}$, the total volume shrinks to $16 \times (\frac{1}{4})^3 = \frac{1}{4}$, but there's still too much overflow. Trying 64 cubes of side length $\frac{1}{8}$ produces a total volume of $64 \times (\frac{1}{8})^3 = \frac{1}{8}$. As we refine the covering, each time doing a better job of approximating the square with little cubes, the volume shrinks successively by a factor of $\frac{1}{2}$. In the limit where the square is exactly covered with cubes, the total volume vanishes.

What the previous exercise tells us is that a square has a finite two-dimensional measure — an area. When we try to determine its one-dimensional measure — its total length — we get infinity. When we try to determine its three-dimensional measure — its total volume — we get zero. A square is two-dimensional, and only a two-dimensional measure — area — gives a finite, nonzero value. This result is actually general. For any geometrical object, measures using too low a dimension lower than the proper value result in infinite values, and measures using too high a dimension result in zero.

What, then, can we infer about measures of fractal objects with noninteger dimensions? Take the Gasket as an example. In Figure 3.13 we attempt to measure the total length of an equilateral Gasket with large side length 1. When we cover just the perimeter with straight lines, we get the answer 3. But there's a lot of Gasket left uncovered. Covering the outer perimeter plus the perimeter of the empty interior region produces a total length of $3 + 3 \times (\frac{1}{2}) = 4.5$. Adding the perimeters of the next smaller empty interiors leads to $3 + 3 \times (\frac{1}{2}) + 9 \times (\frac{1}{4}) = 6.75$. Each time we cover more of the interior, the total length goes up. Furthermore, each time we use lines reduced in length from the previous ones by a factor of $\frac{1}{2}$, there are 3 times more needed than in the previous step in the covering, so the length increases over the previous contribution by a factor of $3/2$. The finer the covering, the more length there is, and this length eventually becomes infinitely large.

Does the same Gasket have area? Figure 3.14 shows a succession of coverings of the Gasket with squares. In the first covering, one square of area equal to 1 covers the entire Gasket. A refined covering with squares whose side length is $\frac{1}{2}$ produces an area measurement of $3 \times (\frac{1}{2})^2 = \frac{3}{4}$. An even finer covering with squares

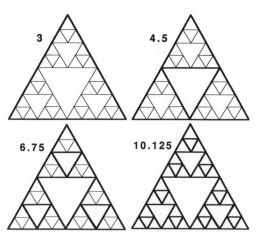

Figure 3.13   *Successively finer measurements of the total length of an equilateral Gasket.*

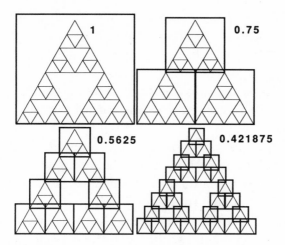

Figure 3.14 *Successively finer measurements of the area of an equilateral Gasket.*

of side length $\frac{1}{4}$ yields an area of $9 \times (\frac{1}{4})^2 = \frac{9}{16}$. The area decreases each time. In fact, in each successive covering with squares whose side length is reduced from the previous side length by a factor of $\frac{1}{2}$, the number of squares increases by a factor of 3, but the area of each decreases by a factor of $\frac{1}{4}$, so the total area decreases each time by factor of $\frac{3}{4}$. Eventually, the total area of such coverings goes to zero. Thus, the Gasket is an object with infinite length (despite being contained within a finite patch of plane), implying dimension higher than 1, but no area— indicating dimension lower than 2. It is unlike any Euclidean object.

An identical conclusion could be obtained for the Koch Curve: zero area, infinite length. At first that doesn't seem reasonable. Surely that wriggly thing we see in Figure 2.10 has finite length. It does. But the wriggly thing in Figure 2.10 is merely a physical realization of the Koch Curve, which is a mathematical, conceptual entity. The conceptual object can be magnified over and over; at each level of magnification, portions of the object are identical to the whole thing. To get all of the length of the Koch Curve—or any mathematical fractal—we have to use smaller and smaller measuring sticks, ultimately *infinitesimally* small, and more and more of them, ultimately *infinitely* many. For the Gasket and the Koch Curve, at least, the sum of these added lengths keep growing, finally without bound. Amazingly, the argument that leads to the conclusion that the length of such objects is infinite can be applied equally well to any of its similar pieces, no matter how small. All have infinite length! That is, the distance along the Koch Curve between any pair of points of the curve is infinite.

The bizarre dimensions of the Gasket, the Koch Curve, and the Cantor Set tell us something about their complexity. Gaskets and Koch Curves writhe around too much to be ordinary, smooth curves, yet not enough to fill in a patch of area. And because the dimension of the Koch Curve is less than that of the Gasket, it is reasonable to conclude that the Koch Curve is closer to a smooth curve and the Gasket closer to an area. That conclusion is certainly consistent with their pictures. The Cantor Set, with dimension between 0 and 1, is intermediate between a collection of unconnected isolated points and pieces of straight lines. At all levels of magnification, the Cantor Set appears to be made of tiny linear stretches, but under closer scrutiny each stretch is seen to break apart. Yet we also can never resolve the individual points; they always cluster too densely.

Dimensions of Physical Objects

What is the dimension of a snaking river, or a snowball, or a circulatory system? The dimension of a self-similar geometrical object can be determined by imagining that the object has mass and by requiring that the mass–length relationship be the same for each of the exact shrunken copies of which the object is made. Natural objects are usually not self-similar. They are more often closer to *self-affine*—that is, made of shrunken, but distorted copies of the whole. Even so, the self-referential character of some natural objects is true only on the average. The fractal attributes of natural objects are usually only statistical properties. The dimensions of natural fractals are not so easily calculated as those in the examples given above. The dimensions of exactly self-affine objects are more difficult to calculate than those of exactly self-similar objects, but in any case, natural fractals typically have no exact symmetry. The mass–length relationships for all the copies in a self-similar fractal are identical. The mass–length relationships for the various copies of a natural fractal differ. In measuring the dimension of a physical object, we have to be content with some sort of average value.

To see how this works, we take a specific example. Figure 3.15 shows three segments of a trace of a photograph of the Mississippi River taken from low Earth orbit. Segment A is the northernmost segment, B connects to the southernmost tip of A, and C connects to the southernmost tip of B. The whole span starts in the north near Cairo, Illinois, and terminates in the south near Natchez, Mississippi. The direct-flight, tip-to-tip distance is about 450 miles; each segment is roughly 150 miles from tip to tip.

If the Mississippi were a self-similar fractal like the Koch Curve, we could identify small, shrunken copies of the whole span repeated throughout it. We could then determine the dimension by identifying N identical copies, each shrunken by a scale factor s, and use the derived result equation (3). But the Mississippi is statistically self-similar. It is made of many pieces that resemble the whole span in a distorted way. In this sense, segment A looks very much like the whole river, *on the average*. In turn, pieces of A look like the whole segment, on the average. We begin by saying that segment A has a "mass" of 1 and a "length" of 1. For convenience (our choice is not unique), we take length to be measured by the side of the smallest square containing A and with sides running north–south and east–west. Next we divide A into N "equal" pieces, each shrunk from the original by some scale factor s. Of course, A is not made of N

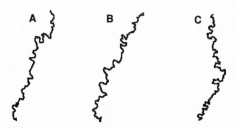

Figure 3.15 *Three segments of the Mississippi River.*

equal pieces of the same size. Therefore, in practice, what we do is find N squares of side length s, oriented in the same manner as the unit-size large square, that contain bits and pieces of A. We'll pretend that the bits contained in each smaller square are pretty much the same. For s too large and N too small, that's likely to be wrong, but we're going to do this a few times — for smaller s and larger N — and then take the average of our results. We hope the averaging smoothes out the errors.

Figure 3.16 shows three successively smaller partitions of segment A, as well as partitions of B and C. The table below contains the relevant data. In each case, the whole segment is enclosed within a smallest bounding square. Then the side length of the partitioning squares is divided by the side length of this bounding square to yield a scale factor s. As you can see, the three data sets are very similar to each other. This corroborates our guess that the three segments are affine copies of the whole span.

We would like to use equation (3) to find the dimension of the river. We could calculate a d value for each N and for each associated $1/s$ and then average them. That would throw away useful information, however — namely, the $N = 1$, $1/s = 1$, values. (Recall that $\log(1) = 0$, so $\log(N)/\log(1/s)$ for this case

Segment	N	$1/s$
A	1	1
	20	12.625
	54	25.250
	108	50.500
B	1	1
	23	12.500
	53	25.000
	119	50.000
C	1	1
	25	12.000
	48	24.000
	124	48.000

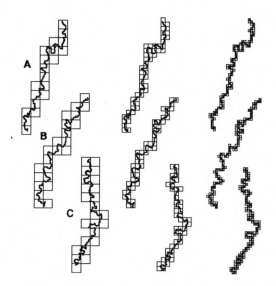

Figure 3.16   *Partitions of the river segments.*

gives 0/0, a ratio indeterminate value.) A better averaging technique uses all the data. Plot $\log(N)$ along the y-axis and $\log(1/s)$ along the x-axis (even the $N = 1$, $1/s = 1$, case). Because, if we do this, equation (3) is equivalent to $y = d \cdot x$, we expect such a plot to produce a straight line with slope d, passing through (or, at least, close to) the origin. Of course, we are dealing with real, not theoretical, data, so we shouldn't be surprised if the data do not fall exactly on a straight line. Instead, we fit the data with a line that comes closest to all of the data. The slope of that "best-fit" line will be the average value of d that we are seeking. All of the data will have participated in determining its value. Figure 3.17 shows a plot of the data associated with segment A. Plots for the other two segments are essentially identical. In each case, we find that the best-fit line has a slope of about 1.2, and we conclude, therefore, that the whole river also has approximately such a dimension. Note too that the data plotted in Figure 3.17 lie pretty close to the best-fit line, suggesting that the various partitions divide the segment into pieces that are very much like each other (on the average).

Our measurement implies that the Mississippi River has a dimension greater than 1. Thus, within the limits imposed by the

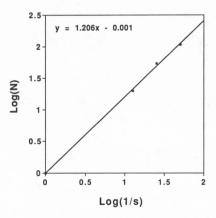

Figure 3.17   *Plot of log(N) versus log(1/s) for the data from Mississippi River segment A.*

resolution of our trace (more on that in a moment), measurements of the river's length are greater and greater as smaller and smaller measuring sticks are used. The coarsest "single-stick" measurement yields 450 miles between Cairo and Natchez, whereas when sticks 1/50th as long (still pretty long) are used (being shorter, they are better able to deal with the tortuous meanderings), a length measurement of about 1000 miles results. (That is, 110–120 sticks, each 1/50th as long as the one that yielded the measurement of 450 miles, are needed to measure its length at this finer resolution.) If the Mississippi were a true fractal (and if we could measure with infinitely fine precision), its length between Cairo and Natchez would emerge as infinite — as would the length of *any* segment between *any* two points! (Recall the Koch Curve.)

Let's look at another way to determine the average dimension of a physical object. This method supposes that we have access to a number of (approximate) replicas of the object of different sizes. Each replica in the collection is assumed to be approximately the same shape and to be made approximately the same way. Again, we take a specific example: this time, balls made by wadding up paper sheets.

We weigh each member of the collection and, for each, measure some characteristic length — say, the longest side length of each irregular paper ball. We do this with real scales and real meter sticks. Such devices read physical units, such as grams and centimeters, not convenient units. To be consistent with our prior discussion, we establish one member of the collection as our "unit" member. For example, we divide the mass and the length of each smaller replica by the mass and the length (respectively) of the largest member of the collection. When we do that, the mass and length of the largest are both equal to 1, the mass of a smaller member is 1/N of the mass of the largest, and the length of a smaller member is reduced by a factor s from that of the largest. The table below gives some typical results.

The entries in this table are determined as follows: Start by cutting from a single sheet of paper (8.5 inches by 11 inches) a sheet $\frac{1}{2}$ as long on each side (4.25 in. by 5.5 in.) and a second sheet $\frac{1}{4}$ as long on each side (2.125 in. by 2.75 in.). Use a single sheet. Tape 4 full sheets together (in a 2×2 arrangement), making a sheet 17 in. by 22 in. Tape 16 sheets together (4×4), making a sheet 34 in. by 44 in. (The less tape you use, the better.) Now wad each of the 5 sheets you have made into tight balls. Try to make them as nearly spherical and as tightly wadded as you can. The larger balls may tend to unravel. If so, use a little tape on the "ears" so that all are nearly equally compact. The largest ball in this set has a mass of 16 sheets, the smallest a mass of $\frac{1}{16}$ of a sheet. Divide the "mass" of each ball into the mass of the largest, 16. (This assumes all sheets weigh the same. You could actually weigh them to get a more accurate result.) This gives, for each smaller ball, the number (N) of identical copies of itself necessary to make up the mass of the largest. Measure the physical diameter of each ball. Divide each by the diameter of

Ball	Mass	N	Diameter	1/s
1	16	1	15.5 cm	1
2	4	4	7.3 cm	2.123
3	1	16	4.6 cm	3.369
4	1/4	64	2.4 cm	6.458
5	1/16	256	1.5 cm	10.333

the largest ball. That gives the scaling factor (s) of the smaller ball relative to the largest. The reciprocal of s is listed in the table, because we want to use the ratio log(N)/log(1/s).

The data are plotted in Figure 3.18. The line best fit to the data yields an average dimension (the slope of the line) of about 2.3 – 2.4. Balls made of wadded paper sheets are natural fractals with dimension between 2 and 3. The attempt to make 3-dimensional objects out of 2-dimensional paper sheets by crumpling and wadding produces structures filled with crinkles. Between crinkles are irregular spaces of different shapes and sizes. Paper balls contain spaces on many different length scales. The bigger the ball, the bigger the possible interior space. A 3-dimensional object, such as a styrofoam ball, might well also have internal holes, but those holes are mostly the same size. Larger 3-dimensional objects have more holes, but the hole sizes are not very different from those found in smaller objects of the same kind. We say that the holes in 3-dimensional objects have a *characteristic size*. The holes inside paper balls made by crumpling flat sheets have no such characteristic size; their statistically self-similar nature causes wadded paper balls to fail to be 3-dimensional.

For both of the examples given in this section, the accuracy of the dimension we calculate depends on our ability to generate a number of replicas that are physically alike but are of widely different sizes. In partitioning the river into large pieces, we run the risk of having one piece especially snaky, another especially

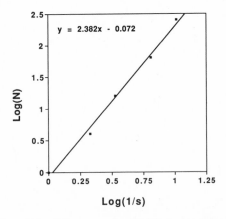

Figure 3.18   *Plot of log(N) versus log(1/s) for paper wads.*

straight. Finer and finer partitions are better, yielding pieces more nearly alike, but we ultimately encounter an operational dilemma: at some point the small squares used in the covering will be the same size as the thickness of our trace. Finer partitions are then meaningless, because they will have lost the ability to discern the way the river wriggles. Because of this, the best we can do is to say that at the level of resolution provided by our trace, the dimension of the Mississippi is about 1.2.

The same situation prevails for paper balls. If a ball is too small, it will fold up without leaving very much interior space. It will be compact and much more like a 3-dimensional object. On the other hand, very large balls are likely to be difficult to crumple with the same packing as smaller ones. Very large, floppy balls are more likely it be closer to 2-dimensional. Thus the dimension of paper balls can be said to be about 2.3 or 2.4, but only for balls between some smallest and some largest sizes. This kind of provisional statement is typical of what can be said for any natural fractal: no natural fractal permits an infinite number of nearly identical replicas.

It is interesting to ponder how our dimension calculation for the Mississippi or for paper balls would change if we were to examine these objects on very different levels of magnification. A mathematical fractal such as the Koch Curve looks precisely the same at all magnifications. We cannot tell whether we are looking at the whole curve or a quarter of it or only a tiny piece. The Koch Curve has the same dimension on all magnification scales; all pieces of it are infinitely long. The Mississippi River and paper balls are not purely mathematical objects, however. Magnifying the river trace a few tens of times would reveal a finite width. At that level, one could see that the two banks would not be parallel; they would wriggle differently, challenging what we meant by "length." The magnified image would certainly be an area with highly irregular edges (where the banks are cut by boulders, fallen trees, sand bars, and tributary streams). A dimension calculation at this resolution would be dominated by the area defined by the surface of the water — and hence would be close to 2, though the curves defined by the water's edges might still have funny dimensions like 1.2. Further magnification would reveal the river's depth; the volume of the river would be unmasked under sufficiently close scrutiny. At this level of resolution its dimension is more like 3.

Similarly, when viewed at a great distance, the paper balls would appear to be pointlike — that is, 0-dimensional. At intermediate view, their approximate 3-dimensional nature would be apparent. Inside, at higher magnification, the sheet surfaces would dominate, leading to a more 2-dimensionality. Ultimately, for the river, the paper balls, and all real objects, the molecules, atoms, electrons, nuclei, . . . would become "seen" in our hypothetical magnifying glass. These pointlike particles suggest that at the highest magnifications, the dimension of all real objects is really essentially 0!

If a real object has no single dimension, in what sense do we dare call it fractal? For natural objects, "fractal" is a metaphor. At some level of resolution, the fractured structure of fractals comes close to the apparently jagged, statistically self-similar character of natural objects. At that level, at least, fractals capture the essence of nature much better than the objects and language of classical geometry. (Recall the Figure on page 3.)

Dimension and Physical Units

When physical units are used, it may be that no object under study has either a unit mass (1 gram, 1 kilogram, 1 pound) or a unit length (1 centimeter, 1 meter, 1 foot). Certainly, it would be an extreme coincidence for an object to have both at the same time. What generalization to our discussion of measuring dimension is required when we use physical units? Our hope, in this case, is to find a relationship between mass and characteristic length for the collection that is analogous to the form found for self-similar geometric objects, namely

$$M = AL^d \qquad (4)$$

In this expression, A is a constant for the whole collection; its value depends on how the objects are constructed and on the system of units used to measure mass and length. The quantity d is the average dimension for the objects making up the collection. A graph of the data in which the logarithm of the mass is plotted along the y-axis and the logarithm of the characteristic length is plotted along the x-axis should still produce a straight line if the data actually satisfy equation (4). To see that, let's take the logarithm of each side of equation (4). We get

$$\log(M) = d \cdot \log(L) + \log(A)$$

which, because $y = \log(M)$ and $x = \log(L)$, is the same as

$$y = d \cdot x + b$$

the equation of a straight line. Note that the dimension we seek is still the slope of the best-fit line, but instead of going through the origin, the line has a nonzero y-intercept, b. In this case $b = \log(A)$. (In our other examples, we always chose one object to have a unit mass and a unit length. When that is true, $\log(A) = \log(1) = 0$.) The coefficient A is 10^b if log base 10 is being used.

We give as an example the data for paper balls again, but now we use physical units. (One 8.5-inch-by-11-inch sheet weighs 4.5 grams; wadded, it has a diameter of 4.6 centimeters.) When plotted this way, the best-fit line has the same slope as before (2.38) but now has an intercept $b = -0.906$. From this we conclude that $A = 10^{-.906} = 0.124$. The units of A are grams/(centimeter)$^{2.38}$. Note that objects with the same dimension can have different A values. For example, rocks and snowballs are probably both 3-dimensional, but a 3-inch-diameter rock weighs a lot more than a 3-inch-diameter snowball; its A value is much larger. When that happens, A is a measure of the mass of the constituent atoms and how tightly those atoms are packed together.

Our goal in this chapter is to give an intuitive feeling for fractal dimensionality. Where precise quantitative values are desired, however, one should clarify exactly which dimension is being measured. Thus our initial operational measurement of dimension focused on how an object's mass varies as smaller and smaller pieces are examined. For obvious reasons, that procedure is said technically to measure the object's *mass dimension*. Using equation (3) to determine the dimension of a self-similar object with N identical pieces shrunken from the whole by scale factor s produces a *similarity dimension*. Snaking the Mississippi through a grid of boxes and using equation (3) on the boxes leads to a *box counting dimension,* or *grid dimension*. There are other dimensional measurement techniques as well. For a self-similar fractal, all such dimensions agree. For fractals that are not self-similar, each technique can produce a somewhat different picture of the structure of the object.

The Importance of Being Fractal

Of what use, you may fairly ask, is the fractal dimension of a natural object? Discussions of fractal dimension can be found in the physical and biological sciences, engineering, medicine, literary analysis, art history, sociology, and many other fields. We present here just a few illustrative examples.

The two examples we described in some detail—the wriggling river and the wadded paper balls—show how dimension and process are intertwined. A young river tends to have a high flow rate and to cut a fairly straight course. As it ages, silt build-up promotes meandering. Tributaries increase in number with the age of the main channel. The dimension of the aging river increases in time (starting from a value near 1). Similarly, the paper wads start as flat, open sheets with dimension near 2. As crumpling begins, large folds are created. With the application of even greater force, large folds themselves become folded, and a range of crinkle sizes results. The ball's dimension climbs steadily during the crumpling. Measurement of a natural object's dimension tells us something about its age. But it may also contain other relic information—if only we were smart enough to see how to extract it—about the dynamical processes that led to the object's geometrical structure. Within the fractal-like properties of the jagged things of nature are clues about the forces and circumstances that made them. Thus the study of natural, fractal geometry is actually the study of evolutionary process. In addition to containing historical information, fractal dimension also tells us something about packing efficiency. A curvelike fractal with dimension between 1 and 2, such as a river, can occupy a small piece of a surface and, at the same time, have a very great length measured along its path. Similarly, the crinkled fractal surface of a paper wad can occupy a small volume but have lots of surface area. Both exhibit economy of packing.

Fractal structures abound in geology on many different scales, presumably as fossil records of complex geological forces applied over many eons. A rather small-scale example is the irregular, statistically self-similar pores found in many rocks. Flows of gases such as radon, and of liquids such as water and oil, are markedly affected by the fractal structure of rock pores. Cracks and fissures are examples of larger-scale fractals, ranging up to many miles long. Crack distributions can strongly influence the pooling and flow of ground water, and their properties are potentially important for developing proper waste management strategies. On the largest scale, across the face of the Earth, are the fractal distributions of all kinds of natural resources: gemstones, precious metals, industrial ores, water, oil. Figure 3.19 shows a

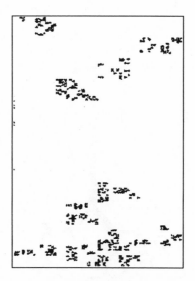

Figure 3.19   *A model of a precious metal distribution many miles on a side.*

simulation of a typical ore lode site distribution. Note that the lode sites are not uniformly distributed; they look similar to a Cantor dust (see Figure 2.06). Developing efficient prospecting techniques would seem to require understanding the clumpy, fractal character of resource distributions. How to measure and interpret the dimensions of these and many other examples of geological fractals is the subject of active research [71].

Not surprisingly, many biological systems have "discovered," through evolutionary luck, the packing economies to be found in fractal structures [44]. For example, lungs are not the empty air sacs pictured in diagrams. They consist of a "tree" of branching pipes (bronchi), growing ever smaller with each branching, and terminating eventually in alveolar sacs. A schematic bronchial tree is depicted in Figure 3.20. Because oxygen molecules are absorbed by red blood cells only where the cells make contact with the walls of the lung, it is an advantage to have as much wall surface area as possible within the volume allowed by the host animal's body. The brachiated, fractal-like structure shown in Figure 3.20 is one solution to this problem, and it is repeated again and again throughout the animal kingdom. Similar packing efficiency requirements have produced branched circulatory systems, serpentine intestinal tracts, and the highly folded architecture of the brain. Again, the issue of just how fractal dimension is related to the form and function of biological structures is being actively investigated.

Another insight into why fractals occur in some geological and biological settings comes from Bernard Sapoval's experiments on

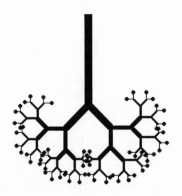

Figure 3.20   *A better representation of a lung.*

drumheads with fractal boundaries [72]. Fractal drumheads behave differently from familiar drumheads, which have Euclidean boundaries. When an ordinary drumhead is struck, the attached membrane establishes large-amplitude vibrations called standing waves. The shape of the standing wave, and hence its sound, are determined by the characteristic length scale of the drumhead surface — the diameter, if it's circular, for example. A drumhead with fractal boundaries has no characteristic length scale, and standing waves are much harder to excite; large-amplitude vibrations of the membrane attached to a fractal boundary die out quickly. (A fractal drum would not be very musical sounding.) Sapoval has speculated that this effect may be responsible for the fractal appearance of coastlines. The erosion of coastlines buffeted by waves creates an irregular, fractal-like shape that, in turn, becomes increasingly effective in damping out wave action. Eventually, such an eroded shape takes on a kind of permanence as it becomes stable against further erosion. Similarly, the fractal form of the circulatory system may have evolved to soften the strong pulsations of the heartbeat. The tendency of fractal surfaces to absorb vibrations may be another mechanism by which they appear around us.

The geometry of growth provides other illustrations of the importance of fractal dimension. Let's do an experiment. Squeeze a very small dab of gel toothpaste onto the middle of a sheet of the clear plastic used to make transparencies for overhead projectors. Place a second sheet on top of the first, and press down as hard as you can so that the toothpaste spreads out into a very thin blob. Now remove your hand and observe what happens around the periphery of the blob. Because of surface tension the toothpaste contracts, allowing air to invade. At first you might expect that the blob would shrink radially into a tighter, rounder blob. But something much less symmetrical and much more interesting happens. The air forms fingerlike intrusions. If you carefully tear the two sheets apart to hasten the intrusion growth, the toothpaste contracts into remarkable stringy structures reminiscent of river networks or, perhaps, lightning bolts. Figure 3.21 is a photograph of the results of such an experiment.

The structures seen in Figure 3.21 are called dendrites because of their fingery nature. Dendrites are often observed when a

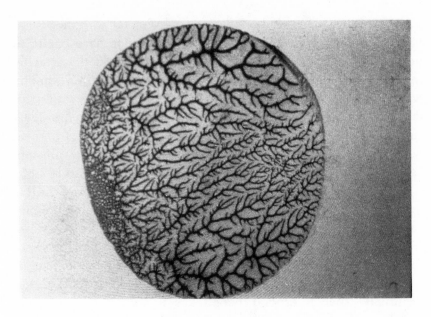

Figure 3.21   *Dendritic structures produced when air infuses into toothpaste.*

lower-density fluid is forced into a higher-density fluid. Although air in toothpaste may be an amusing example, a much more important instance occurs when water is forced into oil. The latter has notable (and bedeviling) practical consequences. A trick that is sometimes employed to try to extract the last drops of crude oil from highly pumped reservoirs is to force water down a central well while pumping out at surrounding wells. Because the water doesn't mix with the oil, as the growing pool spreads, it tends to force the oil that remains into the surrounding pump sites. Unfortunately, this trick is limited by the fractal character of the infusion. Instead of uniformly cleaning out the remnant deposit, the infusing water tends to form fingery paths, leaving significant amounts of untouched oil in the intervening gaps. Experiments showing this effect were first reported in 1951 by W. Engelberts and L. Klinkenberg [73].

The origin of dendritic structures, in some growth processes, at least, can be understood by considering a prototypical phenomenon called *diffusion-limited aggregation*. A diffusion-limited aggregate — a DLA for short — is a cluster of molecules growing within a liquid or solid host. The aggregating molecules wander through the host along erratic paths caused by their accidental collisions with host atoms. The guest molecules are said to diffuse gradually through the host. When the diffusing molecules encounter each other by chance, they stick together. The meandering paths of diffusing molecules lead to clusters that are not very compact. They tend to be dendritic with lots of open gaps, or *fjords,* in between. It's hard for the diffusing molecules to penetrate deep into the fjords, because as they bumble around, they often strike the "walls" of the fjord near where they entered. Figure 3.22 is a photograph of real DLA's formed in sandstone by the precipitation of a copper compound.

Figure 3.23 shows similar structures, grown in two dimensions via computer simulation. The simulation proceeds as follows: A stationary "seed molecule" is enclosed within a circle of arbitrary radius. A "diffusing molecule" is placed randomly somewhere on the periphery of the circle and allowed to execute a *random walk.* In such a walk, the molecule steps, at regular time intervals, a distance that is small compared to the diameter of the circle, either directly north, south, east, or west; the direction of each step is selected randomly. A time-lapse photograph of the

Figure 3.22   *Diffusion-limited precipitates in sandstone.*

Figure 3.23   *Examples of computer-simulated diffusion-limited aggregates.*

resulting motion shows a blurry, irregular, highly overlapped path. If the walk takes the molecule back to the circle, it is replaced by another such molecule somewhere else on the circumference (again chosen randomly). Otherwise, the walk is allowed to continue until the diffusing molecule comes within some critical distance of the seed, at which point it is frozen in place, stuck to the central seed. Figure 3.24 shows six examples of such walks, five of which are "failed" (f) because they reconnect with the bounding circle before sticking (s) to the seed.

The process continues, only now the seed is the two stuck molecules. Eventually, larger and larger clusters grow around the initial seed point. As we indicated before, these clusters are not compact disks; they have open, dendritic shapes. The reason can be seen in Figure 3.25, which shows a walk beginning at point A and heading erratically at first toward the large fjord in the northwest portion of the central cluster. Before entering that fjord, the molecule retraces its steps a bit and then starts heading into the northern fjord. A little way into that fjord, the mole-

Figure 3.24   *Five short random walks entering and leaving the bounding circle (f). One long walk that makes it to the center of the circle (s).*

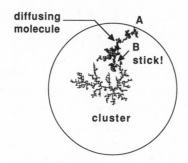

Figure 3.25 *A diffusing molecule hitting a fjord wall before getting deep inside the fjord.*

cule's inebriated walk causes a chance collision with one of the prominent fingers, at point B. There the molecule sticks. The resulting thickening of the fjord wall at B helps make that fjord more nearly impassable for future diffusing molecules. Fjords tend to pinch off, leaving large unfilled regions.

The dimension of DLA's can be measured by surrounding the central seed particle with concentric circles of increasing radii and then "weighing" the mass (actually, counting the number of aggregated moles) inside each circle. As in our previous examples of determining dimension, a plot of log(mass) as a function of log(diameter) should be linear with slope equal to the cluster dimension. (A technical point: You have to be careful to use circles surrounding only the portion of the interior of the clusters that is no longer growing. That's the fractal stuff. The exterior, which is still growing, in general doesn't have the same structure as the insides.) People have grown truly gigantic clusters in the computer, much larger than the tykes of Figure 3.23, and have found that DLA's grown as we described have a dimension of about 1.71.

Structures similar to DLA's occur in a variety of circumstances where the underlying physical causes seem unrelated. Examples include frost patterns on a window pane, deposits on an electrode placed in an electrically conductive solution, "spark" patterns in insulation surrounding high-voltage wires, and, as we've said before, forced infusion of one (low-viscosity—easily flowing) liquid into a second (high-viscosity—gooey) liquid where the two don't ordinarily mix. Why such different physical processes should produce such similar-looking structures is not completely understood, but all seem to have average dimensions in the 1.6–1.7 range.

An important consequence of a DLA's dimension is that the larger it is, the less dense it becomes. Let's be careful about the notion of density. It is a measure not of mass, but rather of how mass is packed. For a flat DLA cluster, a reasonable way to measure density would be to determine the ratio of cluster mass inside a given patch to the area of that patch. Let's say the patch we are using is a disk of diameter D. Then the area of the disk is proportional to D^2, and the cluster mass contained within it is proportional to $D^{1.7}$. Thus the density of a DLA is proportional to the ratio $D^{1.7}/D^2$. As the diameter of the cluster grows, the numerator of this ratio increases less rapidly than the denominator, and the density consequently decreases. A larger DLA is more massive than a smaller one, but it also has more empty fjords, and the fjords grow even more rapidly than the mass as the size increases. In terms of the oil-harvesting trick we discussed, the further a pumping site is from the well down which water is being forced, the more likely it is to be in a fjord and the less likely it is to have oil pushed into it.

The anomalously low densities of fractal clusters may have been necessary for the formation of meteors, comets, planets, and all of the large solid objects in our solar system. It is widely believed that in the earliest days of the solar system, there were no sun or planets; in the earliest days, there was only a huge cloud of gas filled with microscopic dust grains. Later, but still much earlier than the appearance of the Earth, large (though subplanetary) asteroidlike objects (presumably made from the dust) orbited a newly glowing sun (presumably made from some of the gas). Collisions between these "planetesimals" eventually produced the current planets and their moons. But there's an enormous leap that remains unexplained in this path from dust grains to asteroids. Just how did the dust coalesce into rocks? One possibility is a gradual evolution: giant rocks result from the collisions of big rocks, which themselves result from the collisions of medium rocks, and so forth backwards through time. The problem is that at some point in this evolutionary scenario, the largest solids must have been pebbles. And pebbles don't stick together when they collide. They fracture instead. They are too small and their density too high to permit much conversion of the energy of a collision into heat (as occurs when bigger rocks collide).

An alternative view of how large solid bodies came about in the solar system is that dust grains, wafted by breezes in the pre-sun gas cloud, collided gently and stuck together, producing dendritic, hole-filled, fractal dustballs [74]. Like DLA clusters (though they would not be flat), such dustballs would have very low density. (It is estimated that large dust aggregates might have fractal dimension equal to 2.5 or so. Such a dustball of diameter D would have a density proportional to $D^{2.5}/D^3$ — mass in the numerator, *volume* in the denominator.) Two of these putative, highly compressible dustballs would be able to create an even larger and denser dustball if they collided. Giant dustballs (many miles across) might ultimately be transformed into rock through a variety of mechanisms, including melting and crystallization after collision, melting due to the heat of internal radioactivity, and compaction due to the gravitational squeezing of their own mass. This as-yet-speculative view sees the fractal character of dust aggregates as a kind of catalyst for generating solid densities. (It also sharpens the meaning of the biblical adage "from dust, to dust.")

We close this chapter with a few comments about "landscape forgeries." A landscape forgery is a mathematically produced computer image of a seemingly natural scene that does not really exist. An example is shown in Plate 14. This plate is *not* a photograph; it was created using fractal principles: rules designed to make pieces with self-similar properties. A barren, mountainous terrain, for example, can be forged by assigning to every point in a plane a number representing the elevation of the terrain above that point. If these elevations are selected according to some approximately self-affine rules, the image so constructed is also approximately self-affine, the jaggedness of the mountains being similar to that of the hills, the boulders, and the pebbles. The effects of applying different coloring schemes, different lighting methods, and different levels of "water fill" to such images result in extraordinary pictures — sometimes amazingly natural, sometimes discomfortingly unfamiliar. (Similar techniques were used to create the alien world of *Star Trek II: The Wrath of Khan.*) Different elevation assignments lead to different dimensions for these mountain ranges.

The master forger Richard Voss, whom we met in a different context in Chapter One, points out that the most "realistic" mountains have dimensions near 2.2 – 2.3. The fantastic pinnacles of Bryce Canyon have a high fractal dimension, perhaps 2.5 or higher. The dimension of the gentle Blue Ridge Mountains is more like 2.2. Recall that our assessment of the dimension of the Mississippi River was 1.2 or so. Voss's work with clouds and with other geological forgeries [75] (such as the impact crater fields found on the moon, Mercury, and many other bodies in the solar system) suggests a generalization: natural-looking scenes are those whose associated fractal dimension is about 0.2 – 0.3 greater than Euclidean. We evolved in a somewhat mangled, non-Euclidean world and tend to feel comfortable with images that have a familiar dimension. Pictures representing landscapes with dimensions slightly above Euclidean look pretty good to us.

4

Chaos

IN CHAPTER TWO we saw that fractured geometry can be understood in terms of simple evolutionary events. This led us to speculate that the apparent complexity of the often jagged, irregular, and vaguely self-similar objects of the natural world may similarly have simple roots: complexity emerges from processes that occur over and over in an iterative fashion. In this chapter we switch our attention to apparent complexity in time — that is, to sequences of events that contain surprise and unpredictability. Although chance is certainly one source of surprise, it is not essential. Failure of prediction can also arise in circumstances where the future is exactly determined by the past and there is no chance or whimsy. "Deterministic chaos" demands that we reexamine what we can reasonably expect from the enterprise of science; it also holds the promise of enabling us to understand and control our environment in ways not previously imagined. In this chapter we give you some bad news, and a little good news, about chaos. Later, in Chapter Six, you'll get even more good news.

Noise, Noise, Noise

You tune your radio to a weak AM station. Intruding into the programming are hiss and crackle. At a party, you have to shout to your friend because of the loud din in the background. *Noise* is interference with communication, a familiar phenomenon. Communication requires transmission of recognizable sounds or symbols or pictures: some *signal,* some order. Noise diminishes recognition by blurring or distorting the signal and thus producing disorder.

Let's look at a simple pendulum as a kind of prototypical signal source. A simple pendulum is first of all a heavy weight suspended by a long string attached to a fixed support. In equilibrium, the weight hangs vertically downward. When pulled to one side and released, the pendulum swings back and forth along a smooth arc, repeating its motion over and over. But a simple pendulum is more: it swings in a plane with an amplitude (the maximum displacement from equilibrium) that is small compared to its length, and it has negligible friction. Clearly, a simple pendulum is an idealization, but real pendula often come pretty close to conforming to the ideal. What is the pendulum's signal?

Figure 4.1 shows a representation of a swinging pendulum and a way of extracting information from its motion. Imagine that the weight of the pendulum is a pot of ink. There is a small hole in the bottom of the inkpot, and small drops leak from the hole, falling on a strip of tape below the pot. The tape is pulled at a constant speed in a direction perpendicular to the arc of the swinging pendulum. The drops that hit the tape make a record of where the inkpot is from instant to instant.

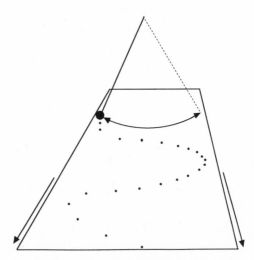

Figure 4.1   *As it swings, a pendulum drops ink on a tape.*

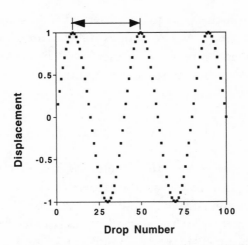

Figure 4.2   *A record of the motion of a simple pendulum.*

If we set the tape upright, the trace of drops undulates smoothly like that shown in Figure 4.2. The width of the trace corresponds to the pendulum's position in the arc of its motion. The length of the tape can be converted into time, because it is being pulled at a constant rate. For example, one drop might hit the tape every 0.1 second. As the drops are made to fall faster and faster, the dots on the tape get closer and closer together. A continuous stream of ink paints a continuous curve on the tape. Such a smooth, snaking curve is called a *sine curve*. The motion of the pendulum is said to be periodic (as is its sine curve graph). That is, the motion exactly repeats itself after the passage of a characteristic time called the *period of the motion*. The time designated by the double-arrowed line in Figure 4.2 is one period.

In the seventeenth century, Isaac Newton formulated three laws that describe how one body's motion is altered by the pushes and pulls (forces) exerted on it by other bodies in its surroundings. Newton also developed a mathematics of continuous change called calculus. The two wedded together constitute the discipline of *physical dynamics*, whose goal is to predict where a body will be and how it will be moving, given the forces acting on it and some data on initial position and motion. Physical dynamics is a model, a representation, not the motion itself. Nonetheless, in those circumstances where the forces are relatively simple, dynamics often yields predictions that agree well with the motion that is actually observed. The simple pendulum is one such instance. Not only does Newtonian dynamics predict periodic oscillatory motion—describable by a sine curve—it also predicts that the period of the motion can be expressed as

$$T = 2\pi\sqrt{\frac{L}{g}}$$

where L is the length of the pendulum (the distance from the fixed support point to the center of the weight hanging from the string) and g is the acceleration experienced by all bodies in free fall near the surface of the Earth because of the Earth's gravitational pull (about 32 feet per second per second (32 ft/sec²) or, equivalently, about 9.8 m/sec²). Note that the predicted period of the pendulum does not depend on the weight's mass, on its chemical composition, on the amplitude of the motion, on the air pressure or humidity, or on a whole bunch of other plausible parameters. It depends only on the length of the pendulum and the strength of gravity. (The experimental discovery that a simple pendulum's period is independent of its amplitude is attributed to Galileo, who, in one version of the event, is alleged to have timed the oscillations of the great chandelier in the cathedral of Pisa with his pulse during Sunday services. However charming, this story is apparently apocryphal, because the chandelier was not purchased until after Galileo's death. [76])

To see how close a real pendulum comes to the prediction of dynamics, consider Figure 4.3. The plot shows data taken with a pendulum close to 1 meter (m) in length. The amplitude of the motion started at about 5 cm and gradually became less as a result of the effects of friction. But the period is remarkably stable and has a value extremely close to that predicted for a simple pendulum 1 m long. The pendulum is symbolic of the success of Newtonian dynamics. Systems like the pendulum, in which simple forces act, often have regular and predictable motion. These systems include (among many possible examples) automobiles, airplanes, and deep space probes. Much of the technology of the developed world, along with its associated standard of living, is based on such systems.

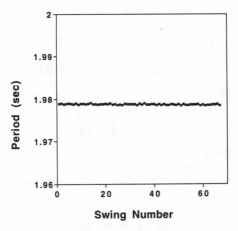

Figure 4.3   *A plot of the period of a pendulum versus time.*

The Newtonian vision of the universe is a clockwork: in it the future is determined by the present, and the present by the past, in a perfectly predictable fashion. Reality is often at odds with this vision, however. Surprises, accidents, and matters of apparent chance are commonplace. And there is noise. Consider Figure 4.4, which shows daily mean temperatures in Albany, New York, for the year 1991 [77]. It is reasonable to expect seasonal variations in the temperatures at Albany — highs in the summer, lows in the winter. To a first approximation, that is what Figure 4.4 reveals. Seasonal undulation is the signal in the data displayed. But significant daily variability is also quite evident.

Erratic fluctuations in signals are much more the norm than the exception. A few additional examples are depicted in Figures 4.5 – 4.7. (In all of these figures, the data are the dots. Lines connecting the dots are drawn to help you visualize the sequence of events. The lines do not represent actual information on a finer scale.) Figure 4.5 presents the daily weight of milk produced by cows on a dairy farm, also near Albany, during 1991 [78]. There is a vague seasonal variation signal in these data, but the noise is impressive.

Figure 4.6 shows the daily closing values of the Dow Jones Industrial Average Index (DJI) for 1991, extracted from *The Wall Street Journal*. Viewed over a long enough interval (years), the Dow Jones Index is observed to increase gradually. But wildly unpre-

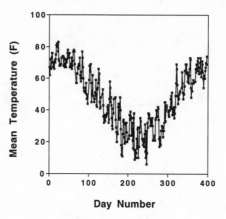

Figure 4.4 *Daily mean temperatures in Albany, New York, in 1991.*

dictable rises and falls are often superimposed on that upward trend.

When the intervals between heart beats for a well, resting heart are measured, one finds some scatter. The well heart is not exactly a ticking clock. (Actually, there is evidence that heart rates tend to become more regular with age; a very regular heart may be symptomatic of the onset of disease.) But when a sick heart undergoes fibrillation, all manner of irregularity breaks out. Figure 4.7 is a record of the intervals between heart beats for a patient experiencing atrial fibrillation [79].

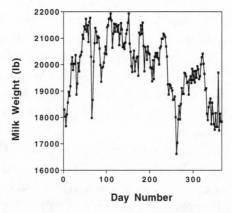

Figure 4.5 *Daily milk weights from cows on a dairy farm in 1991.*

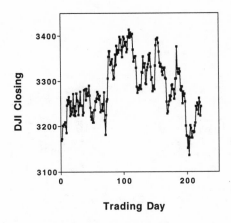

Figure 4.6   *Daily Dow Jones closings for 1991.*

Is the universe partitioned into two kinds of phenomena, one regular and predictable, the other filled with irregular fluctuations? Though this is a tempting hypothesis, it is not borne out by observation. The apparent regularity of the pendulum we saw in Figure 4.3 is illusory. We just didn't look closely enough. When measured with a very precise timer, the period of the pendulum is also seen to be a noisy signal! Figure 4.8 shows the same data as in Figure 4.3, but with the vertical scale magnified by a factor of 50. At this resolution we can make out very subtle irregularities. Thus not even the paradigm clockwork mechanism is free of

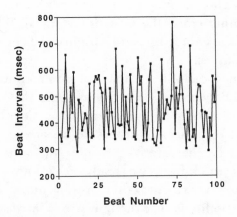

Figure 4.7   *Atrial fibrillation.*

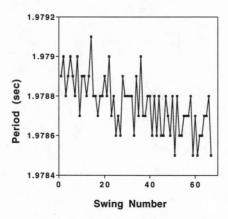

Figure 4.8   *The period of the pendulum of Figure 4.3 examined under high precision.*

noise. (Note that the data seem to have a downward trend in Figure 4.8. Friction causes the pendulum to have an ever smaller amplitude of motion, while also making it repeat faster — about 0.000005 sec [5 microseconds] per oscillation. Measured at this high precision, the pendulum's period is detected to depend weakly on its amplitude. Galileo's well — though surely irregular — pulse could never have revealed such a delicate effect.)

On the Roots of Chaos

Some signals harbor so little noise that we have to work very hard to find it. In other cases, the noise *is* the signal. Where does this irregularity in the course of events originate? There are a number of plausible sources.

Certainly one possibility is that we don't know all the rules that govern the behavior of those systems that defy accurate prediction. It wouldn't be surprising for a model based on incomplete ideas to be lacking in predictive power. Indeed, ignorance undoubtedly characterizes the current status of many areas of inquiry: all too often, we really *don't* know all the rules (though we hope we're always uncovering more of them).

Another possibility is that the universe is intrinsically random; that is, at some primitive level there aren't any rules. Contrary to

Einstein's famous statement of faith in a completely knowable universe, maybe God actually *does* play dice! The best evidence we have at the moment strongly suggests that in the microrealm — in the realm of the atom — chance is essential. There, random occurrence *is the rule*. Perhaps, in ways we don't yet understand, the fundamental stochasticity (randomness) of the atom somehow infects the macroscopic world. Why automobiles and airplanes would be less susceptible to this infection from "below" than the stock market or a sick heart makes rigid adherence to this point of view pretty difficult to defend, however.

Even if all the rules are strictly deterministic and are known exactly, there is still room for the intrusion of disarray when many variables are involved. An accurate description of a system consisting of a large number of independent pieces, despite all of them behaving perfectly deterministically by simple rules, can require more information than we can process. For example, consider a deck of 52 ordinary playing cards. Suppose a "shuffle" is known to send the top card to the 17th place, the second card to the 32nd place, the third card to the 4th place, etc. — some complicated, but perfectly well-defined set of rules. Suppose the initial state of the deck is ace of spades on top, king of spades next, . . . , three of diamonds next, two of diamonds last: the order found in a freshly opened box. Now apply the "shuffle" rules a few times. Very soon the scrambled cards are so lacking in mnemonic clues that we give up trying to keep track of the deck's state. We often say that eventually the order of the cards becomes "randomized," though in fact, chance may play no role at all. What we really mean is that the complexity of the system overwhelms our ability to make accurate predictions; its behavior is as good as random.

Although it is certainly true that all matter is made up of extremely large numbers of pieces, these pieces often behave collectively, yielding new properties and a reduced set of parameters. The 10^{22} or so atoms in a diamond cooperate to make a rigid, sparkling crystal. The 10^{24} or so atoms in a beaker of water cooperate to make a wet liquid that quenches your thirst. The 10^{27} or so atoms of the air in a room cooperate to transmit sound waves. In a similar fashion, the 10^{10} or so neurons in the brain (metaphoric brain atoms) cooperate to create thought, and the

10^8 or so persons (metaphoric social atoms) in the United States cooperate to form a (more-or-less) functional society. Bulk properties of matter are very different from the properties of the constituent atoms from which the matter is made. The collective behavior of a system of many pieces can frequently be described by only a few variables — many fewer than when the individual pieces act independently.

Unless we state otherwise, we will be dealing with just these kinds of systems; we assume that the many constituent pieces always work together and that the number of relevant variables is always small. But even in these instances — that is, where there are few relevant variables and where the rules are exactly known — unpredictability can emerge in the form of deterministic chaos. Let's see how.

Linear Dynamics

We now digress briefly from our pursuit of chaos to establish a few conventions and learn a useful graphical procedure. We will be dealing with dynamical systems. A *dynamical system* is a set of related phenomena that change in time in a deterministic way. By determinism we mean that the future of the system is completely determined by its past. The ingredients of a dynamical system are (1) a definition of the state of the system and (2) a rule for change called the dynamic. (Recall that physical dynamics refers to the study of change in a body's motion as a result of applied forces. Here, dynamics has a more general meaning. It refers to the study of all kinds of deterministic change.)

Let's return, for a concrete example, to the first iterative process we considered in Chapter One, audio feedback. Look at Figure 1.1. The *state* in the audio feedback dynamical system is the loudness of the sound emitted by the speaker. The *dynamic* is the following rule: In each pass through the amplifier, multiply the loudness by the "effective gain." Thus if the sound level starts at 60 decibels (a unit of loudness) and the effective gain is 1.5, then after one pass through the amplifier the loudness will be $(1.5)(60) = 90$ decibels, and after a second pass it will be

$(1.5)(90) = 135$ decibels. We can write a general formula for audio feedback dynamics:

$$L_{n+1} = g \cdot L_n \tag{1}$$

where L_n is the loudness after n passes through the amplifier, and g is the effective gain. The effective gain in each pass through the amplifier is $g = (1) + (\text{additions}) - (\text{losses})$, where "additions" means the fractional increase in loudness due to energy input from the circuit, and "losses" means the fractional decrease in loudness due to energy losses of all kinds. When additions equal losses, the loudness is unchanging ($g = 1$). Note that g is greater than 1 when more energy is put in in each pass than is lost; g is less than 1 when the reverse is true.

Exactly the same relationship as that shown in equation (1) characterizes other familiar situations. For example, a simple interest payment scheme at a bank is a dynamical system. The state is your balance, and the dynamic is the following rule: After each year, multiply by the quantity (1 + the effective interest rate). Say the effective interest rate is 5% and you have $100.00 (in 1993 dollars) in your account. Then after one year you will have $(1.05)(\$100.00) = \$105.00$ in 1994 dollars (use 1.05 because the decimal equivalent of 5% is 0.05). And after two years you will have $(1.05)(\$105.00) = \$110.25$ in 1994 dollars. The effective interest rate is the stated interest rate minus the inflation rate, and the calculation gives effective value measured in the dollars with which you started your account. If inflation just matches the stated interest rate, your account maintains constant value. For your account to grow in value, the stated interest rate has to exceed the rate of inflation. Letting B_n be the value of your balance after year n (measured in year-zero dollars) and letting r be (1) + (stated rate) − (inflation rate), we have the dynamical relationship

$$B_{n+1} = r \cdot B_n$$

which, as we said, is formally the same as equation (1).

Another, perhaps familiar, example that has the same formal dynamics is the growth of a colony of single-cell organisms. Let's assume that these organisms reproduce at regular intervals, called generations, and that they have fixed birth and death rates. Let P_n be the number of cells in the population after the nth generation. Let f be (1) + (birth rate) − (death rate). (Birth rate − death rate is called the fecundity of the population.) Then

$$P_{n+1} = f \cdot P_n$$

determines the population size generation after generation.

A typical question related to linear dynamics is "Given some starting state, what is the state after some number of iterations?" For example, what is the value of your bank account after 17 years if you start (in year zero) with $217.36 and the effective interest rate is 6.3%? (Similar questions could equally well be posed for loudness and population size.) A straightforward approach to this kind of problem is to multiply $217.36 by 1.063, then to multiply that result by 1.063, then to multiply that result by 1.063, and so on, 17 times:

$$B_0 = \$217.36$$
$$B_1 = (1.063)B_0 = (1.063)(\$217.36)$$
$$B_2 = (1.063)B_1 = (1.063)(1.063)B_0$$
$$= (1.063)^2 B_0 = (1.063)^2(\$217.36)$$

$$\cdot \ \cdot \ \cdot$$

$$B_{17} = (1.063)^{17}B_0 = (1.063)^{17}(\$217.36)$$

The latter result is easily evaluated with the help of a hand calculator; the answer is $614.11. At this effective interest rate, the value of your account (in year-zero dollars) would have almost tripled in 17 years. The bank gives you your initial deposit back plus twice that amount in interest payments. Note that 6.3% per year times 17 years is a little over 100%. In other words, if the bank gave you effectively 6.3% of your initial deposit each year, and you didn't allow the interest to draw

interest, you would have received total interest payments equal-
ing only your initial deposit. Interest accruing interest is called
compounding, and the resulting growth can be surprisingly rapid.
Note also that the conclusion that your initial deposit triples in
value is independent of the actual starting value. Finally, note
that if you started with the special value $B_0 = \$0.00$, your ac-
count would never change under this set of compounding rules,
no matter how long you waited.

Although such calculations are relatively easy to do for simple,
fixed-rate, compounding problems, they become tedious for
more elaborate and realistic dynamics. Often we are content to
make quick, qualitatively correct estimates of how the state of
our dynamical system changes rather than working through the
details of finding exact numerical results. Such estimates are
facilitated by using graphs. Here's how.

For the sake of specificity, let's stick with the language of the
banking version of the dynamics we just described. Make a plot
of the relationship $B_{n+1} = r \cdot B_n$ with B_{n+1}, the new balance, on
the y-axis and B_n, the old balance, on the x-axis. In terms of x
and y, the relationship $B_{n+1} = r \cdot B_n$ is equivalent to $y = r \cdot x$,
which you recognize as a straight line—called the compounding
line—with slope equal to r and with y-intercept equal to 0. An
example in which r is greater than 1 is shown in Figure 4.9. Now
suppose you know the starting account value is B_0 and you want
to find the next five values. Well, B_1 is $r \cdot B_0$. That's the y-coordinate of a point on the compounding line directly above B_0. Again,
see Figure 4.9. So far, so good.

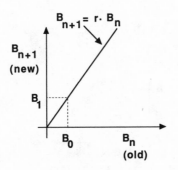

Figure 4.9   *Graphical plot of the relationship* $B_{n+1} = r \cdot B_n$.

What about B_2? That value is the y-coordinate of a point on the compounding line directly above B_1. How do we establish where B_1 is on the x-axis? A neat trick is to draw on the same plot a second line, through the origin, with slope equal to 1. The x- and y-coordinates of every point on that line are equal to each other (that is, new balance = old balance everywhere along this line). See Figure 4.10. This new line, with equation $B_{n+1} = B_n$, enables us to locate B_1 quickly on the x-axis: we just draw a horizontal line from B_1 on the y-axis until it hits the line $B_{n+1} = B_n$. The coordinates of the point of intersection are both B_1. To find B_2, we draw a vertical line up from B_1 on the x-axis until it hits the compounding line; that point of intersection has y-coordinate equal to B_2. And so on. Actually, Figure 4.10 has more lines than necessary. A neater and easier construction is shown in Figure 4.11.

The sequence of dotted lines in Figure 4.11 traces out the *graphical iteration* of the dynamical system $B_{n+1} = r \cdot B_n$. In graphical iteration, the interesting information can be read off the $B_{n+1} = B_n$ (new = old) line — the 45° line. Graphical iteration consists of (a) establishing your starting value on the 45° line [the point (B_0, B_0)], (b) drawing a vertical line that hits the compounding line, and (c) drawing a horizontal line from that intersection point until you hit the 45° line again. The latter intersection point has coordinates (B_1, B_1). Then you iterate, using (B_1, B_1) as the start of the next step in the construction. One full iteration consists of drawing one vertical line from the 45° line to the dynamics line and drawing a second, horizontal

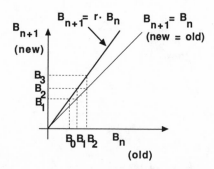

Figure 4.10   *Plot of $B_{n+1} = r \cdot B_n$ along with the line $B_{n+1} = B_n$.*

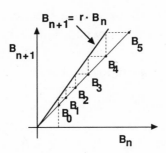

Figure 4.11 *Five successive iterates of* $B_{n+1} = r \cdot B_n$.

line from there to the 45° line. In the figure, the compounding line is completely above the 45° line, so iterating forward in time proceeds by going *up* to the compounding line and then to the right to the 45° line. There is no other option.

Undoubtedly, you are content with the idea that the value of your account will grow as time goes on under the fixed interest rate rule described in the preceding figures, provided that the actual interest rate exceeds the inflation rate—that is, r is greater than 1. This kind of constant rate compounding leads to what is known as *exponential growth*. Each year, as your account grows, the amount the bank gives you grows as well. See Figure 4.12. The difference between B_0 and B_1 is smaller than that between B_1 and B_2, which in turn is smaller than that between B_2 and B_3, and so on. The increment in the value of your account is greater each successive compounding. The sequence of values B_0, B_1, B_2, . . . is said to constitute a *time series*. The graph on the right in Figure 4.12 depicts this time series; it plots the succes-

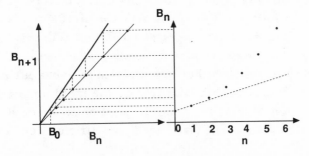

Figure 4.12 *The time series for the dynamical system* $B_{n+1} = r \cdot B_n$.

sive B_n values above the respective times (n) at which they occurred. The line drawn between B_0 and B_1 on the time series graph serves to emphasize that the balance grows more rapidly than linearly (in linear growth, B_2, B_3, B_4, . . . would all lie on that same line).

Unfettered exponential growth, though never realized in exact detail in nature, is sometimes observed as an approximation to reality. The amplification of the sound from a speaker in an audio feedback loop grows roughly exponentially until the energy requirements of keeping the sound growing become too high for the electronics to meet. A population of cells grows roughly exponentially until the population becomes so large that its environment cannot supply enough nutrient or disperse enough of the colony's waste products to maintain the growth. Please note that the time series for exponential growth—the plot of B_n versus n—is *not* linear, despite the fact that the rule for generating exponential growth—$B_{n+1} = r \cdot B_n$—*is* linear. A nonlinear time series does not imply nonlinear governing rules. (The simple pendulum is another example of a dynamical system that is governed by linear rules but whose time series, the undulating sine curve, is not linear.)

We note, parenthetically, that the results of exponential growth can often be amazing. Here's an illustrative fiction: A meteorite falls to Earth in the middle of the United States, carrying a tiny mold colony. This mold takes to Earth conditions and doubles its area each day. The colony starts as a one-square-millimeter blob. How long will it take the mold to cover the entire surface of the 48 contiguous states? To appreciate what is involved, you first have to recognize that the 48 contiguous states have an area of some 10^{19} square millimeters! That is, the mold has to increase its area by a factor of 10^{19} if it's to cover them all. That seems like a pretty tall order. Will this happen in any reasonable time? At first it seems that it would take many centuries to accomplish. But in fact, it takes only a little over 63 days! Why? Well, each day the area of the mold doubles, so in n days the colony's area is 2^n times its starting area. Set 2^n equal to 10^{19} and solve for n. (We need our friend the logarithm again.)

$$\log(2^n) = \log(10^{19})$$
$$n \cdot \log(2) = 19 \cdot \log(10)$$
$$n = 19 \cdot \log(10)/\log(2) = 63.1 \text{ days}$$

Question: How many days does it take the colony to cover half of this area? Another question: How many days would it take the colony to cover the whole Earth—whose surface area is 5×10^{20} mm^2, 50 times that of the contiguous states?

In each of the three examples of linear dynamics that we've given, the behavior of the state of the system is governed by a single control parameter, either g (effective gain), or r (interest rate), or f (fecundity). Like the gain on the amplifier, you can think of the system's control parameter as a "knob setting." What kind of time series do we get when the control knob setting in each of these examples is less than 1? In these situations, losses outweigh additions in the amplifier circuit, inflation outstrips the stated interest rate at the bank, and deaths outnumber births in the cell population. Figure 4.13 shows the answer. You can see that when losses predominate, the time series heads inexorably toward zero—no sound, no bank balance, no population. Note that when we do the graphical iteration in this case, the vertical lines have to go downward instead of up, and the horizontal lines have to go to the left instead of to the right. When the dynamics line is below the 45° line, there is no other choice. The time series associated with Figure 4.13 is drawn in Figure 4.14. Such a time series is said to display *exponential decay*.

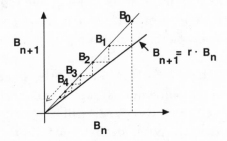

Figure 4.13   *Iteration of $B_{n+1} = r \cdot B_n$ when r is less than 1.*

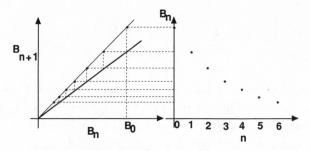

Figure 4.14   *The time series for exponential decay.*

In linear dynamics the state at the end of n iterations (compoundings, passes, or generations) is the beginning state multiplied by $(1 + \text{effective rate})^n$. In the latter expression, "effective rate" means the fractional change in the state in one iteration as a result of additions and subtractions. Now suppose that the frequency of updating the state value is increased; that is, suppose each of the prior iteration intervals is made up of N subintervals. For example, instead of compounding every year, your bank compounds every quarter ($N = 4$) or every month ($N = 12$) or every day ($N = 365$). In this case, the proper expression for the state value at the end of n of the larger intervals (such as n years) is the initial state value multiplied by $(1 + [\text{effective rate}]/N)^{nN}$. The proper effective rate is the rate per larger interval divided by the number of subintervals, and the proper number of iterations is nN.

Here's a specific example: The effective rate at your bank is 5% per year, but interest is compounded quarterly. How much is an initial deposit of $100 worth at the end of 2 years? If interest were compounded annually, the answer would be $(\$100) \cdot (1 + 0.05)^2 = \$110.25$. Compounding quarterly produces $(\$100) \cdot (1 + [0.05]/4)^8 = \$110.45$, a slightly higher yield. Compounding daily at the same annual rate gives $(\$100) \cdot (1 + [0.05]/365)^{730} = \$110.52$ — slightly higher still.

In the limit that iteration becomes continuous — that is, N approaches infinity — [effective rate]/N becomes vanishingly small, but the exponent nN becomes infinitely large. The limit actually produces a finite result that can be written

$$(1 + [\text{effective rate}]/N)^{nN} \rightarrow e^{[\text{effective rate}]n}$$

where the quantity e is a fundamental constant called the exponential constant, whose value is 2.718281828459. . . . If your bank continuously compounded your interest, in the example given, after 2 years your account would be worth $(\$100) \cdot (e^{[0.005] \cdot 2}) = (\$100) \cdot (e^{0.01}) = \$110.52$. (In this example, daily compounding is almost as good as continuous compounding. The two results do not differ until the sixth decimal place, so you would have seen the difference only if you had had an initial deposit of $1,000,000.) Many calculators have an e^x button; you punch in x, then e^x, to obtain the desired result.

Here's a bit more language that is commonly used in dynamical systems analysis to describe the things we have just been observing. Because the state of the system — the loudness, bank balance, or population size — is changed only after a finite interval according to the rules we defined, these dynamical systems are called *discrete*. (There are also *continuous* dynamical systems, the pendulum being one example.) The special state 0 (0 decibels, $0, 0 population) is said to be a stationary state or, more often, a *fixed point*. Application of the dynamical rules to a fixed point does not change its value. In the case where the slope of the dynamics line (that is, the control parameter) is less than 1 (Figure 4.13), 0 is said to be a *stable* fixed point because any small amount added to 0 decays back to 0 in short order. In slightly dramatic fashion, 0 for this case is also said to be an *attractor* (all states near it are attracted to 0 as the dynamics are played out). The stable fixed point is similar to the bottom of the bowl represented in the upper part of Figure 4.15. A marble at the bottom of the bowl is in equilibrium (left). When it is displaced up the side of the bowl a bit and then is let go, it slides back down toward the bottom (right).

Conversely, when the slope of the compounding line is greater than 1 (Figure 4.12), 0 is an *unstable* fixed point — a *repellor*. A repellor is similar to the situation shown in the lower part of Figure 4.15. There the bowl is turned upside down, and the

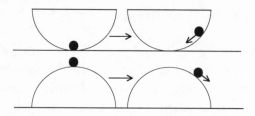

Figure 4.15   *Two types of mechanical equilibria: stable (upper) and unstable (lower).*

marble is placed very carefully at the top (left). The marble is in equilibrium in this carefully arranged state, but any slight jiggle causes it to career away from the equilibrium point (right). Obviously, something special happens to the dynamics when the slope is just equal to 1 (that is, when losses and additions just balance out). For that special value all starting states are stationary; nothing ever changes. The behavior of the system is said to undergo a *bifurcation* at the critical value of control parameter equal to 1; the qualitative character of the dynamics changes from having one stable fixed point to having one unstable fixed point (or the other way around).

Limits to Growth

Nowhere in nature are infinite noises, infinite bank balances, and infinite populations observed. The growth, which in Figure 4.12 threatens to progress without bound, is always limited by finite resources: finite energy, finite wealth, finite nutrient. Thus linear dynamics is a poor representation of reality for more than just the beginnings of growth. If we want a better approximation to the way nature works, we must modify our linear models: we need *nonlinearity*.

One simple method for producing limited growth is to add a second rule. For example, we might allow exponential growth up to a certain critical state value and then apply a different, though still linear, dynamic beyond that value. Figure 4.16 shows what we mean. Because we started discussing dynamics by using interest compounding as our primary example, we continue that language here. But keep in mind that we're not talking about

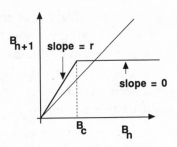

Figure 4.16 *A two-rule dynamical system.*

realistic banks; we're talking about the general behavior of simple nonlinear dynamics. In any case, the "bank" in the figure allows your account to grow exponentially for balances below a critical value, B_c. Once your balance exceeds B_c, however, a new rule goes into effect. This new rule is also linear, but its slope is 0 (and its y-intercept is not 0).

What is the effect of such a second rule? The answer can be seen in Figure 4.17. Both small and large starting balances have the same fate: the dynamics eventually carries all balances (except $0, of course) to the value designated B_f in the figure, where it then stays. The balance B_f is a new, and stable, fixed point. It occurs where the second dynamics line intersects the 45° line. (The fixed value $0 also occurs where the dynamics line intersects the 45° line. Because the slope of the dynamics line is greater than 1 there, $0 is unstable—it repels.) You can see that generally, for any dynamical system, where the dynamics line (or curve) crosses the new = old, 45° line, the point of crossing has to be a fixed point. If you try graphical iteration starting at such a point, your attempt to go vertically to the dynamics line keeps

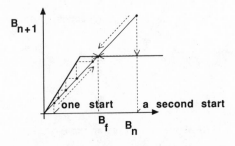

Figure 4.17 *Iteration with the dynamics of Figure 4.16.*

you right where you are. The condition that the second dynamics line has a slope equal to 0 is another special case of a more general result. If the second dynamics line has *any* slope less than +1 and greater than −1, the point where it intersects the new = old, 45° line will be a stable, attracting, fixed point. You can easily convince yourself of this by drawing a few pictures, as in Figure 4.18.

An interesting thing happens when the slope of the second dynamics line is exactly −1. See Figure 4.19. As soon as the iteration carries past the second fixed point, the dynamics locks into a cycle that repeats itself every second iteration. Such a periodic behavior is called a 2-cycle. (In this nomenclature, a fixed point could be called a 1-cycle, because it repeats every iteration.)

What happens when the slope of the dynamics line at the second fixed point is more negative than −1? Figure 4.20 provides the answer. In this case *both* the fixed point at $0 and the second fixed point are unstable. Like two bumpers on a pinball machine, both rebuff any attempt of the dynamics to settle down. As we will see, the mutual repulsion of the two fixed points creates havoc.

Fixed points of dynamical systems made of multiple linear pieces occur where the dynamics lines cross the 45° line. Graphical iteration suggests that any such fixed point is stable when the slope of the dynamics line is less than 1 in magnitude (when we consider magnitude alone, we throw away any minus sign) and

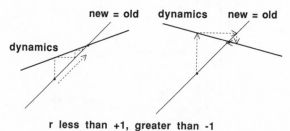

Figure 4.18   *Stability of the second fixed point.*

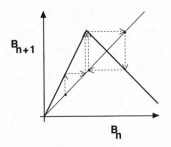

Figure 4.19 *When the dynamics line has slope -1, you get periodic cycling.*

that it is unstable when the slope is greater than 1 in magnitude. When the slope of the dynamics line equals 1 or -1, a bifurcation in the qualitative behavior of the system occurs. We can see in more comprehensive terms why all of this is true.

What happens to a starting balance near the fixed value, B_f? Let the starting value in either Figure 4.18 or 4.19 be written as $B_0 = B_f + d$. What is the first iterate, B_1? The most general form of a linear dynamics rule can be expressed as

$$B_{n+1} = r \cdot B_n + b \tag{2}$$

In banking terms, such a rule corresponds to "Multiply previous balance times (1 + effective interest rate) and then give a bonus b (or subtract a fixed fee, if b is negative)." Apply the governing equation (2) to B_0:

$$B_1 = r \cdot B_0 + b = r \cdot (B_f + d) + b = r \cdot B_f + b + r \cdot d$$

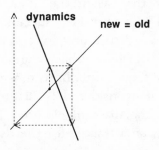

Figure 4.20 *When the dynamics line has slope more negative than -1, the second fixed point is unstable.*

(The second equality is true because of the assumed value of B_0, third by a simple rearrangement of terms.) Now we use the fact that B_f doesn't change when the dynamics is applied; that is, $B_f = r \cdot B_f + b$. We can reexpress B_1 as

$$B_1 = B_f + r \cdot d$$

Proceed to find B_2:

$$B_2 = r \cdot B_1 + b = r \cdot (B_f + r \cdot d) + b = r \cdot B_f + b + r^2 d = B_f + r^2 d$$

Clearly, repeated application of the dynamics leads to this general result:

$$B_n = B_f + r^n d$$

If the magnitude of r is greater than 1, each successive iterate will be farther from B_f than the last; the fixed point in this case is therefore unstable. If r has magnitude less than 1, each successive iterate will be closer to B_f than the last; the fixed point in this case is therefore stable. If $r = 1$, then $B_0 = B_1 = B_2 = \ldots$, no changes; if $r = -1$, however, $B_0 = -B_1 = B_2 = -B_3 = \ldots$, a 2-cycle results. Although our discussion assumed a starting balance a little greater than the fixed balance, everything we said is also true for starting balances a little less than B_f.

Note that the linear compounding rule $B_{n+1} = r \cdot B_n + b$ is an affine iterated function system (Chapter Two): r is a scaling transformation and b is a translation. This IFS consists of a single (invertible) rule. The state being transformed is a single number —such as a balance—not, as in our prior discussion of IFS, a picture. Remember that when the affine IFS was contractive (scalings less than 1 in magnitude), successive application of the transformation rules led to a unique, compact picture. So, too, for the banking IFS: when r has magnitude less than 1, B_n converges to a finite fixed value. In turn, we can also say that the deterministic affine IFS considered in Chapter Two is a dynamical system operating in the space of pictures. The picture to which the IFS converges is an attractor for the dynamics.

The Tent Map

The tent map is the simplest dynamical system harboring deterministic chaos. It is closely related to the limited-growth dynamics we discussed in the previous section. The tent map's definition is as follows: The state is a single variable, x_n, which takes on values ranging from 0 to 1. The dynamical rule is

$$x_{n+1} = s \cdot x_n \quad \text{if } x_n \text{ is less than } 0.5$$

and $$\qquad\quad = s \cdot (1 - x_n) \quad \text{otherwise} \qquad\qquad (2)$$

Here's how the rule works numerically. Suppose $s = 1.5$ and $x_0 = 0.3$. Under these conditions x_0 is less than 0.5, so $x_1 = (1.5) \cdot (0.3) = 0.45$; x_1 is still less than 0.5, so $x_2 = (1.5) \cdot (0.45) = 0.675$; x_2 is now greater than 0.5, so $x_3 = (1.5) \cdot (1 - 0.675) = 0.4875$. And so on. For qualitative purposes, graphical iteration is quicker.

The graph of the dynamics appears in Figure 4.21 for an example in which s is greater than 1. The (unit) square box drawn around the dynamics lines is not part of the dynamics. It serves to remind us that all of the state values are contained between 0 and 1. Above every value of x_n that we could use as a starting value (between 0 and 1) is the value of x_{n+1} (in bold) that we get by applying once the rule given as equation (2). The figure clearly shows why the word *tent* is associated with this dynamical system. The word *map* is meant to suggest that a unique future state

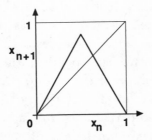

Figure 4.21   *A tent map for s greater than 1.*

corresponds to each past state; the past is "mapped" into the future.

We do not claim that the tent map is a very realistic model of anything in nature. It is interesting primarily for what it can teach us about the origins of deterministic chaos. Nonetheless, it does have some connection with the population dynamics of single-cell organisms living in a finite environment on the surface of a Petri dish. Suppose P_{max} is the maximum number of cells that the environment can support. Let $x_n = P_n/P_{max}$ be the fraction of the possible viable cell sites that is actually occupied. And suppose that when x_n is small, the population is governed by simple linear dynamics:

$$x_{n+1} = x_n + b \cdot x_n - d \cdot x_n$$

This relationship says that in generation $n + 1$, the (fractional) population is that in generation n, plus additions due to births ($b \cdot x_n$), minus losses due to deaths ($d \cdot x_n$). When the population is large, there will almost certainly be cells essentially everywhere ready to "give birth" to new cells. But suppose the population's birth rate depends on "free space." If x_n is the fraction of viable space occupied, then $1 - x_n$ is the fraction still available, and the effective birth rate can be given approximately by $b \cdot (1 - x_n)$. In that case,

$$x_{n+1} = x_n + b \cdot (1 - x_n) - d \cdot x_n$$

The tent map follows from these two dynamics equations if (a) we say that a population is "small" when x_n is less than $\frac{1}{2}$ and that it is "large" otherwise, and (b) we measure time in units of $1/d$. That is, we rescale time so that d is equal to 1. The tent map's parameter, s, can then be understood as the population's birth rate parameter measured in these rescaled time units.

From a purely formal perspective, the tent map's control parameter, s, is the slope of the line to the left of $x_n = \frac{1}{2}$, rising out of the origin. The line to the right of $\frac{1}{2}$ has a slope equal to $-s$. At $x_n = \frac{1}{2}$, the two parts of the rule agree; there, $x_{n+1} = s/2$. That value is the maximum height of the tent. (Thus you can think of the height of the tent as the control parameter.) To ensure that x_n stays between 0 and 1, we have to require that s not be negative and that it not be larger than +2. In the latter case, the tent would pop through the unit square, allowing x_n to get outside of the interval [0, 1].

When s is between 0 and 1, the maximum height of the tent is less than $\frac{1}{2}$—the top of the tent lies below the point $(\frac{1}{2}, \frac{1}{2})$. For s-values less than 1, the tent intersects the 45° line at only one point, the origin. Thus, for this case, $x_n = 0$ is the only fixed point, and, because the slope of the dynamics line (that is, s) is less than 1 there, 0 is an attractor. (See also Figure 4.22.) When s is greater than 1, the tent pops through the 45° line and a second fixed point appears. The fixed point at 0 is unstable when s is greater than 1 (the slope of the dynamics line there is too positive), but so is the new fixed point (the slope of the dynamics line there, −s, is too negative). Here is a situation in which x is confined between 0 and 1 but both of the fixed points in this range repel. What, one wonders, does the time series for this case look like?

Figure 4.23 shows a few iterations of the tent map for a case where s is greater than 1. The associated time series (to the right) doesn't look very orderly, but this may be a result of starting with a poor value of x. We really need the help of a computer to generate lots of data. Three time series—each 200 entries long and each starting at x = 0.5, but for different s-values—are plotted in Figures 4.24–4.26. We start with the top of the tent almost touching the top of the unit square, s = 1.9999. (The

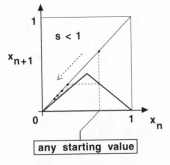

Figure 4.22   *When s is less than 1, 0 is an attractor.*

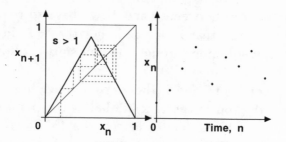

Figure 4.23   *A few iterations when s is greater than 1.*

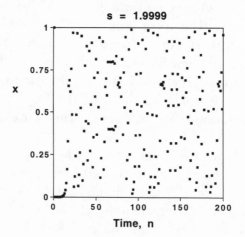

Figure 4.24   *Time series for the tent map with s = 1.9999.*

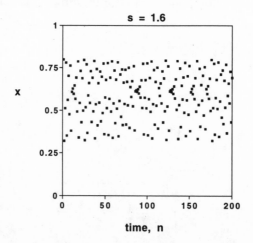

Figure 4.25   *Time series for the tent map with s = 1.6.*

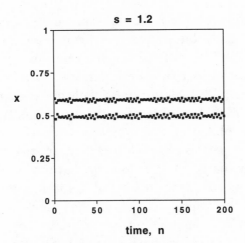

Figure 4.26   *Time series for the tent map with s = 1.2.*

reason for picking such an unlovely value for s is that if we had picked s exactly equal to 2, the first iterate of $x_0 = 0.5$ would have been $x_1 = 2(0.5) = 1$, and the second would have been $x_2 = 2(1 - 1) = 0$. All successive iterates of 0 remain 0, because that is a fixed point. We wouldn't have learned much.) The time series sprawls all over the plot, without any apparent organization.

On the other hand, when s is set to 1.6, as in Figure 4.25, the still helter-skelter time series is much more compressed. The range of x-values appears to be bounded between about 0.3 and 0.8. When s is reduced further, to 1.2, as in Figure 4.26, a remarkable time series results: two narrow, though slightly fuzzy bands of values. For this case, the slope of the dynamics line at the second fixed point is getting close to -1. We mentioned before that when the slope of the dynamics line at a fixed point is exactly -1, a periodic 2-cycle occurs about that point. In Figure 4.26, the value of x_n hops back and forth between two bands in an *almost* periodic fashion, except that successive values never exactly repeat. In fact, when we examine them in fine detail, we find that none of the time series produced by the tent map for s-values between 1 and 2 ever exactly repeats! This non-repeating, seemingly erratic behavior is called deterministic chaos.

It may seem that the phrase *deterministic chaos* is an oxymoron. *Chaos* colloquially means "without order," whereas *deterministic* means "the future is rigidly determined by the past." Thus deterministic chaos sounds like "predictable unpredictability" or "orderly disorder." Well, technically deterministic chaos refers to the irregular output of a deterministic system. According to Robert Devaney [80], deterministic chaos is characterized by three properties: (1) sensitive dependence on initial conditions, (2) lots of unstable order, and (3) mixing. We will say more about the first two properties soon. "Mixing" means that a chaotic time series will eventually (though you may have to wait a long time) come as close as you please to any value within the allowed range of the output. In Chapter Six we will show how the properties of unstable order and mixing can be exploited for practical purposes. The language of modern science is filled with words that have coexisting commonplace and technical meanings (work, momentum, charm, strangeness, and so on). Unfortunately, the two meanings are sometimes only vaguely related. Thus the first-time reader of science has to be vigilant in order to keep the common and the technical usages straight. Throughout the rest of this book, we use *chaos* and *deterministic chaos* interchangeably, and both have the technical meaning expressed by Devaney in terms of the three properties cited above.

Can the tent map accommodate periodic behavior? We can begin to answer this question by examining the plots shown in Figures 4.27–4.29. (Unless explicitly labeled otherwise, all maps operate on state values between 0 and 1.)

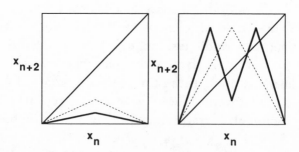

Figure 4.27   *The second-return plot for the tent map.*

In Figure 4.27 we plot x_{n+2} versus x_n (this is called a second-return map). That is, above every value of x_n, we plot the value that would come up after *two* iterations of the tent map. We show two cases: s less than 1 (left) and s greater than 1 (right). The bold lines correspond to all such (x_n, x_{n+2}) pairs. For reference, we have also traced in the "ghost" of all (x_n, x_{n+1}) pairs, which is the tent map itself, for the two cases. The second-return map for s less than 1 is another tent, but squatter than the "first-return" tent (the dotted "ghost"). The value of x where the bold, second-return tent intersects the 45° line is a value (0) that repeats itself every two iterations. It coincides with where the first-return tent intersects the 45° line: 0 repeats itself every time *and* every other time. For s less than 1, the tent map has no 2-cycles.

As s is made larger than 1, the second-return map—the bold structure—develops a pucker in the middle. It is no longer a tent, but more like an M. The places where the "M" crosses the 45° line, in Figure 4.27, locate values of x that repeat themselves every *second* time when plugged into the tent recipe. There are four such crossings. Two of the crossings are identical to where the first-return "ghost" also crosses. Those values of x (the two [unstable] values of the fixed points of the tent map) repeat themselves *every* time and, hence, repeat every other time as well. But what do we make of the other two crossings of the "M"? Figure 4.28 tells us. Figure 4.28 is x_{n+1} versus x_n again, but with a "ghost" of the x_{n+2} versus x_n "M" in the background for reference. Start at one of the "M" crossings—say, the one designated by a. Graphical iteration takes that first value to the value designated by b after one iteration; b is the second "M"

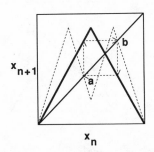

Figure 4.28   *The 2-cycle of the tent map.*

crossing. A second iteration takes us back to a. In other words, the values of x_n at a and b don't repeat every time; they repeat every other time. These two crossings of the "M" pinpoint the values of x_n (for the given value of the tent parameter s) that form a 2-cycle (repeat every two iterations).

Perhaps a numerical example would be useful. Let s = 2. The two values of x that constitute the 2-cycle are $\frac{2}{5}$ and $\frac{4}{5}$. To see this, start at $x_0 = \frac{2}{5}$. Then according to the tent rule,

$$x_1 = s \cdot x_0 = 2 \cdot \tfrac{2}{5} = \tfrac{4}{5}$$

and

$$x_2 = s \cdot (1 - x_1) = 2 \cdot (1 - \tfrac{4}{5}) = \tfrac{2}{5} = x_0$$

Thus this cycle repeats every two iterations.

So we see that when s is greater than 1, the tent map harbors a 2-cycle behavior. But is that behavior stable? Consider Figure 4.29. There, a starting value of x is chosen close to one of the values (a) that repeat every two iterates. Successive iterations starting as shown rapidly get farther and farther from the two cyclic values (at a and b). This instability is characteristic of the 2-cycles of the tent map for all s-values. A neater way of saying the same thing is to look at the second-return map, x_{n+2} versus x_n. See Figure 4.30. Pick a starting value of x_n near either of the 2-cycle values. One iteration on that plot of either starting value (x_n) takes us to values (x_{n+2}) that are still farther from the

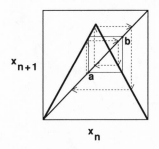

Figure 4.29 *The 2-cycle is unstable.*

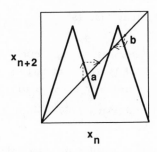

Figure 4.30   *Iterations on the second-return map.*

2-cycle values. If the 2-cycle were stable (attracting), such iterations would bring us closer. (Every second x would get closer to the special, repeating values if the 2-cycle were stable.) Iteration carries us away from the 2-cycle values, because the slopes of the dynamics lines at those points are steeper in magnitude than 1.

We can continue our quest for periodicity by asking about 3-cycles, 4-cycles, ... The third-return map for the tent is shown in Figure 4.31 for $s = 2$. It cuts the $45°$ line in eight places. Two of those correspond to where the first-return map (the "ghost") cuts the $45°$ line: values of x that repeat every time also repeat every three times. The other six values correspond to *two* 3-cycles. The slopes of the dynamics lines at these intersection points are all steeper in magnitude than 1, so both 3-cycles are unstable.

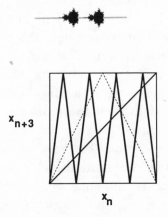

Figure 4.31   *The third-return map for the tent, $s = 2$.*

Here are three general results:

1. Each successively higher-return map cuts the $45°$ line once — at 0 — for s less than 1. That is, the tent has no periodicity other than that of the stable fixed point for s less than 1.

2. For s greater than 1, each successively higher-return map has more and more "M-like crinkles."

3. Where these crinkled dynamics intersect the $45°$ line, the slopes are all greater than 1 in magnitude.

Thus the tent map permits plenty of periodic behaviors. In fact, as s approaches 2, cycles of all periods are permitted. But for s greater than 1, all of these cycles are unstable!

The Butterfly Effect

Deterministic chaos was brought to the fore of our scientific awareness by happenstance in the early 1960s by the meteorologist Edward Lorenz [81]. (Actually, many of the properties of chaos had been worked out prior to Lorenz's discovery by Henri Poincaré [82] in the nineteenth century, by George Birkhoff [83] in the 1920s, by Mary Cartwright and J.E. Littlewood [84] in the 1940s, and by Stephen Smale [85] in the late 1950s. The word *chaos* was first used in a technical context by T.Y. Li and James Yorke [86] in 1975.) Lorenz was attempting to model convection in the atmosphere by computer calculations. Convection comes about when the air temperature near the ground is higher than that in the upper atmosphere. The warm, low-altitude air rises and is replaced by falling cool air from above. As the rising air cools and the falling air warms, a churning flow is established. If conditions are just right, and there are no horizontal winds to scramble them up, convection churning can produce very beautiful, ordered cloud patterns like those shown in Figure 4.32. If conditions are a little different, the churning can destroy these patterns. Lorenz was investigating the irregularities in such convection flows, using some nonlinear equations as a model. After completing one long simulation, he decided to repeat the calculation starting at a time T, partway through the simulation, taking

Figure 4.32   *Convection rolls in the atmosphere.*

as the starting state for the second run the data generated for time T in the first run. The data he input came directly from the computer's printout: three digits of precision instead of the six the computer used internally for calculations. Amazingly, the new calculation and the computer's printout of the prior run began to disagree noticeably in a short time; eventually they became totally unrelated. Lorenz's nonlinear weather dynamics was behaving chaotically, and the slight initial discrepancy between his two calculations was driven by that chaos into total doubt.

Figure 4.33 shows how this effect is manifest in the tent map with s near 2. In this figure, we see how two states evolve by iteration with the tent map for a fairly large value of its control parameter. The tent dynamics tends to stretch the distance between the states and to fold them over (that is, iteration via the righthand portion of the tent tends to flip the order of two nearby states). In a short time, states that were initially nearby have little relation to each other.

A quantitative example of the explosive divergence of initially close states is shown in Figure 4.34. The graph depicts the

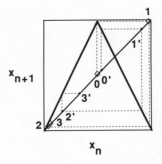

Figure 4.33 *Two states, initially close, diverge under iteration of the tent map for s > 1.*

evolution of the difference between the nearby starting states — $x_0 = 0.5000$ and $x'_0 = 0.5001$ — for $s = 1.99$. For the first 10 iterations or so, the difference between the two states remains small, but quickly thereafter the difference between the states grows to almost the maximum possible value (+1 or −1), and the ordering of the states fluctuates wildly (sometimes the difference is positive, other times negative).

Differences can grow in linear dynamics, too, but in a much more controlled fashion. Consider the linear dynamics rule $x_{n+1} = (1.99) \cdot x_n$. After one iteration, the state 0.5000 grows to

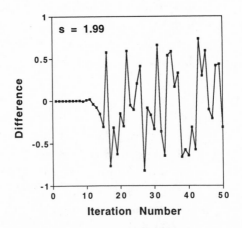

Figure 4.34 *Propagation of difference via the tent map for s near 2. The difference between two states that were initially only 0.0001 apart stays small for only a few iterations.*

(1.99)·(0.5000) and the state 0.5001 grows to (1.99)·(0.5001). Their initial difference — 0.0001 — grows to (1.99)·(0.0001). On the other hand, the difference between the two states *relative* to the size of one of them — the 0.5000 state, for example — is initially 0.0001/0.5000 = 0.0002; after one iteration, the relative difference is (1.99)·(0.0001)/([1.99]·[0.5000]), still exactly 0.0002. It is easy to continue this argument, with the conclusion that the relative difference remains constant. In other words, although it is true that the difference between two states can grow under linear dynamics, the growth is much tamer than that for nonlinear, chaotic dynamics: the order of the states is preserved, and the relative difference remains unchanged.

Figure 4.34 contains a profoundly disturbing implication. No measurement of any physical, biological, or social state is infinitely precise; there is always some uncertainty. If the underlying dynamics governing the evolution of the measured state is chaotic (albeit deterministic), the initial uncertainty (however small) will inevitably grow so large that long-range prediction becomes impossible! When chaos is present, the future is dramatically altered by the subtlest of changes in the start-up state. Such exquisite fussiness explains why deterministic chaos manifests "a sensitive dependence on initial conditions." Somewhat more poetically, Lorenz has referred to this sensitivity as the Butterfly Effect, which suggests that even the most gentle, unaccounted-for perturbation (such as the seemingly irrelevant flapping of a butterfly's wings somewhere out of sight) can produce, in short order, abject failure of prediction.

The dilemma for forecasters is not solely founded on the uncertainty of measurement. No computer can store and manipulate numbers with infinite precision. Thus no computer can calculate long, chaotic time series with great accuracy. Furthermore, different computers (including hand calculators) treat arithmetic slightly differently. Thus two different types of computers that are set to the task of calculating the same chaotic time series (same control parameters, same starting values) eventually produce results that have little in common with each other. Indeed, time series that are calculated on a single machine starting with the same initial data, but using different — though mathematically equivalent — algorithms, also diverge after a suf-

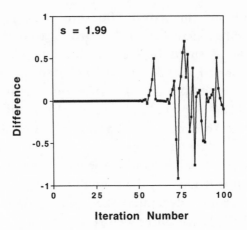

Figure 4.35   *Comparison of two equivalent algorithms for computing tent map output. Starting with the same initial data, the two agree to high precision for only about the first 50 iterations.*

ficient number of iterations. That annoying and confounding realization is demonstrated in Figure 4.35.

A form of the tent map that is mathematically equivalent to the conditional equation (2) is

$$x_{n+1} = s \cdot (0.5 - |x_n - 0.5|)$$

where the vertical lines surrounding $x_n - 0.5$ mean "take the absolute value of" — that is, whether the difference is negative or positive, make it positive. (You should try a few numerical examples to convince yourself that our claim of equivalence is true.) Computers treat the conditional equation and this absolute-value form slightly differently. They yield results that can differ in the last digit. Figure 4.35 was generated by using numbers with 16-digit precision and by taking $s = 1.99$ and $x_0 = 0.5$ for both tent map forms. At first, the two outcomes agree to within 15 digits. Subsequently, the differences are amplified by the nature of the chaotic iteration. Eventually, the two calculations show no relationship to each other. Which is correct? The answer is neither, because both use finite precision.

In view of the pessimistic results shown in Figures 4.34 and 4.35, we may be forced to reexamine our expectations for the scientific enterprise. We are accustomed to judging the worth of a scientific theory by its ability to synthesize data and organize observations *and* by its ability to make accurate predictions. Does the existence of sensitive dependence on initial conditions mean that the science of chaotic systems is doomed to be purely taxonomic, devoid of predictive power?

Strange Attractors

Fortunately, deterministic chaos does allow some prediction. First, Figures 4.34 and 4.35 imply that chaotic systems permit predictability in the short term. But long-term, statistical predictions are also possible for chaotic dynamics. That is, it is possible to make such statements as, "In the next 100 iterations, the following approximate state values are more likely to occur than others: . . . ." Though they are less exact than one might expect for deterministic time series, such predictions are potentially valuable nonetheless.

The basis for the claim of statistical prediction originates in the observation that the dynamics allows only certain ranges of outcomes. Consider Figure 4.36. Look at the lefthand tent first. Point a corresponds to the top of the tent. Point b is the value to

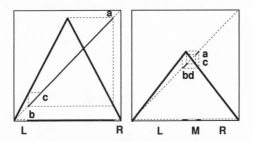

Figure 4.36 *Two tent maps showing successive iterations starting from the maximum value allowed by the dynamics (values designated by "a" in each case). For both situations, graphical iteration starting either below b or above a quickly enters the region between b and a and then stays within that region. This region contains the attractor of the chaotic dynamics. On the left hand side, the attractor covers the entire interval from b to a. On the right hand side, the attractor is divided into two separated pieces—from b to d and from c to a.*

which point a maps, and point c is the value to which point b maps. For the lefthand case, c lies below the nonzero fixed point. Imagine starting with a value in the dotted region L. Quickly the action of the tent dynamics takes that value into the bold region bounded by b on the left and by a on the right. (Try it.) Once the state value enters the bold region, it stays within, because to get out would require achieving a value greater than the maximum permitted by the dynamics (a). As the height of the tent is lowered, an interesting thing happens. That is the case on the righthand side of the figure. There point a maps to b, and b maps to c, but c is above the nonzero fixed point. One more iteration, to d, reveals that there is an empty region between the intervals b to d and c to a that is not accessible to the dynamics. The only way to get in the middle region, M, is to have started within it. Figure 4.37 magnifies the relevant portion of the picture. In this case, all starting values eventually become trapped within the two bold intervals. Successive iterations hop back and forth between the bands but never reemerge in a dotted region. This trapping behavior is also demonstrated in Figure 4.38, where three different starting values ultimately lead to two-band hopping.

The experiment shown in Figure 4.38 seems to suggest that there is a kind of sameness about nonrepetitive, chaotic time series if we wait long enough for start-up behavior to settle down. That notion is buttressed by the results of Figure 4.39. To help make sense of the chaotic time series associated with the tent map when s is greater than 1, we use histograms to keep

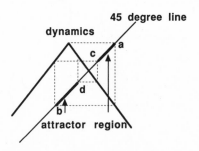

Figure 4.37 *A magnification of the attractor of the tent map on the right hand side of Figure 4.36.*

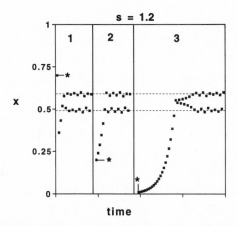

Figure 4.38   *Three tent map time series, all associated with the same parameter value (s = 1.2). The series start at different values (indicated by the "stars"), but quickly take on similar behaviors.*

track of the frequency with which x-values get visited by the dynamics.

(A reminder about histograms may be in order. Histograms are a crude but useful statistical measure; they provide a convenient way of showing the preference for certain values in a large

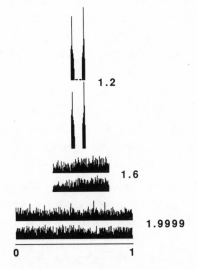

Figure 4.39   *Three pairs of histograms for tent map time series. Each pair displays behavior starting at different initial values but with the same parameter value.*

collection of outcomes. Suppose you want to see what is the most common height among your friends. It is likely that if the height measurements were made *extremely* precisely, no two of your friends would share exactly the same height. (In fact, at that level of precision you would find that a given friend's height would not be constant but would vary during the day.) Thus a plot of the number of friends at a given height as a function of very precise height wouldn't be very revealing: there would be a lot of zeroes and an occasional one. On the other hand, too coarse a measure of height — such as "somewhere between 0 and 10 feet," for example — wouldn't be very edifying either, because *all* your friends would fall into such a coarse interval. To make a useful histogram, you must choose the proper "bin" size. In the height example, binning your friends together in height intervals of 1 inch — or 2 inches, or 6 inches — might provide useful information. Making the "correct" choice is something of an art.)

In Figure 4.39, the bin size is one pixel on a computer display. The entire x-range (0 to 1) spans 220 pixels. Thus the interval associated with each bin is $1/220 = 0.00454$. . . . The length of the line above each bin is a measure of the total number of times during the period of data collection that the value of x falls within the range of that particular bin. In each of the specific cases shown, the top histogram is for $x_0 = 0.1$, the bottom for $x_0 = 0.5$. In each case, the same number of data points are plotted for each run. To minimize the effects of start-up choice, data are plotted starting with the 1001th iteration.

The striking similarity of the two histograms for each s-value suggests that the chaotic output of the tent map actually has underlying regularity. Despite our computer's finite precision, and despite the necessary consequence that its output is "wrong" iteration after iteration, not just any old outcome appears. For each s, all sequences of x_n-values that we compute eventually dance around within the dynamically allowed intervals in approximately the same way. Inside each dynamically allowed interval lies an attractor on which the chaotic ballet unfolds. Unlike isolated stable fixed points, though, these chaotic attractors consist of densely clustered points, often with fractal properties. For these reasons they are said to be "strange."

The Family Portrait of Attractors

When we perform an actual computation of a chaotic time series from some starting value, round-off errors grow and grow, and the series becomes very dissimilar to the theoretical, infinite-precision one. At the same time, there is an underlying attractor trying to tame the error growth. What the computation produces then is a sequence of values never exactly on the attractor, but always "shadowing" it. Two close starting values spawn two sequences that rapidly diverge. On the other hand, each sequence shadows the attractor, approaching it closely iteration after iteration. Eventually *any* such computation can be used to estimate the likelihood that a time series will visit the attractor's different regions. Despite its strict determinism, chaos is most aptly characterized by statements of likelihood.

We wonder what the family of attractors of the tent map looks like. To find out, we proceed as follows: We make a plot of some number of approximate values of x on the attractor (vertically) for each allowed value of control parameter, s (horizontally). Pick an s-value. Start the calculation anywhere: $x_0 = \frac{1}{2}$ is convenient. Run the calculation for some number of iterations to get rid of start-ups that are not near the attractor. Then plot along a vertical column enough subsequent x's to get an idea of what the attractor looks like for that value of s. Move to the next s and repeat. A typical result for the tent map is shown in Figure 4.40. To make this figure, 300 successive x-values were calculated for

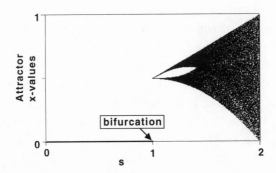

Figure 4.40   *The bifurcation diagram for the tent map.*

every value of s between 0 and 2; 150 x's, starting with iteration number 151, were plotted vertically above each respective s. The plot shown has the following features: For s-values between 0 and 1, the attractor is the single fixed point 0. At $s = 1$ a bifurcation occurs: a new qualitative behavior, a nonzero attractor, emerges. (Accordingly, the plot is called a *bifurcation diagram*.) As s increases from 1, the nonzero attractor appears to start as a single point that then quickly splits into two values — a 2-cycle. But magnifying this part of the plot many times shows that the sharp 2-cycle really is two narrow bands of values. Eventually the bands thicken and merge (at $s = \sqrt{2}$. Do you see why?) and for higher s-values, only one, ever-broadening band is observed. As s approaches 2, the attractor spreads out and occupies the entire x-interval between 0 and 1. The graininess of the diagram beyond the band merging arises from the relatively small number of x-values plotted for each s. If we had used 1500 values instead of 150, the graininess would have been much less apparent.

Such is the portrait of chaos in the tent map.

Some Dust in the Tent

Fractal geometry and chaotic dynamics are intimately intertwined. That connection will be more evident in the next chapter, but here we can show at least one — already familiar — fractal connected to the tent map.

When the top of the tent is less than 1, all x-values are trapped between 0 and 1. What happens when we take the bold step of allowing the top of the tent to be greater than 1 — that is, of pushing the vertex of the tent through the top of the unit square (s > 2)? The situation is shown in Figure 4.41. You can see that any x lying below the little triangular piece defined by the top of the unit square and the portion of the text sticking out above it will eventually reach $-\infty$. To convince yourself that this is true, trace out what happens to an x inside of the central "bad" zone in Figure 4.41. To do this requires extending the sides of the tent

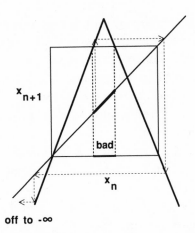

Figure 4.41   *The tent map with control parameter greater than 2.*

below the unit square on each side and extending the 45° line. An x inside the central "bad" zone has a first iterate greater than 1. The second iterate goes negative and, once there, heads tragically off to $-\infty$ with dispatch.

But there are other bad zones besides the one in the middle. You could start outside the middle bad zone and, after one iteration, get into it. All of the values of x mapping into the bad zone in one iteration are shown in Figure 4.42. After one iteration the story is the same—a tragic end. In turn, these zones will be hit after one iteration by starting values in other bad zones. . . . In fact, one might well ask whether there are any "good" x's at all—that is, any points that don't wind up, after a while, heading off toward a tragic outcome.

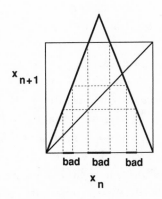

Figure 4.42   *More bad zones.*

The answer is that there are quite a few. Some that don't wind up at $-\infty$ are those that sit directly under the vertical dotted lines in Figure 4.42. Those x's are okay because they eventually iterate to 1, which then iterates to 0—one of the two fixed points. Others that are good are the nonzero fixed point and all x's that iterate to it. (We're thinking theoretically, not computationally, now.) There are still more. We know that the tent map has all kinds of periodic (though unstable) cycles. All those special x-values and the x's that (theoretically) iterate to them are "good." The set of all such good values is disconnected but pretty dense. Indeed, this collection of good points is a Cantor dust, like those we encountered in Chapter Two.

Fact and Fiction

The fact that deterministic processes are able to generate seemingly random output has profound practical and philosophical implications. As a natural consequence, there is widespread interest in the properties of chaos. Many thousands of technical articles have been written on the subject, and some of the language and ideas of chaos are regularly referred to in popular magazines, in the daily newspaper, in novels, plays, and movies, and on TV shows. Some users of the language of chaos are disciplined and critical; others are more casual, sometimes creating images and impressions that are not supported by the facts.

It is useful, in closing this chapter, to remind ourselves of what chaos is and of what it is not. Chaos is irregular output from a deterministic source. The future of a chaotic behavior is completely determined by its past. Chaos is not chance or randomness. What makes chaos confounding is the way measurement uncertainties expand. If you are dealing with a chaotic system and you start with an incompletely specified initial state, that initial uncertainty grows rapidly and makes exact prediction in the long term impossible. Nonetheless, statements clarifying the limits and likelihoods of future behaviors can still be made for a chaotic process.

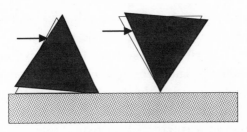

Figure 4.43 *Two systems initially in equilibrium respond differently to small perturbations. On the left, the system is in stable equilibrium; on the right, the system is in unstable equilibrium. In neither case is there any doubt what will happen in the future.*

The growth of uncertainty in chaotic phenomena *is* the Butterfly Effect. In colloquial usage, however, the Butterfly Effect is often taken to mean that minuscule influences can *cause* enormous outcomes (for example, the beating of a butterfly's wings in Brazil can cause a tornado in Texas). Well, one can think of plenty of examples of small influences leading to large effects. For example, Figure 4.43 shows two triangular wedges. Both are initially in equilibrium. The one on the left, sitting on its base, is in *stable* equilibrium; the one on the right, perched precariously on its tip, is in *unstable* equilibrium. When the one on the left is given a gentle tap, it may rock a bit, but it eventually returns to its initial state. A gentle tap delivered to the one on the right, however, causes that one to topple over. In the first case, a small perturbation dies out with no net effect. In the second case, a small perturbation leads to a very large outcome. In general, when a system is nearly in stable equilibrium, small influences quickly disappear. When a system is far from stable equilibrium — as in the case of the wedge teetering on its tip — small influences quickly amplify and a new state emerges. In both of the examples given, though, the net effect of the small influence is completely predictable. Neither is an example of the Butterfly Effect.

In order for the Butterfly Effect to be relevant, we need chaos: the state of the system has to be continually changing. Chaos is associated with systems evolving far from equilibrium (either stable or unstable) — systems that are capable of swapping energy and matter with their surroundings. The interplay with the surroundings keeps these systems from settling down. We reemphasize that the kinds of systems we are talking about are composed of many pieces acting collectively. The number of relevant variables needed to describe such systems is many times smaller than the number of pieces, and although the pieces interact over short distances, the relevant variables tend to have much longer length scales. Underlying the various possible outcomes of such chaotic systems are strange attractors. It is reasonable to argue that the attractor is determined by a few large-scale control parameters, but the actual state at any instant depends on how we got started. In this view, the tornado in Texas is a state on the attractor; it is in the future of (almost) any starting state. At first, the state of the atmosphere with the butterfly beating its wings in Brazil is very close to the state with the butterfly at rest. Eventually, as the dynamics unfolds, these two states diverge. The tornado is going to happen sometime in both scenarios. In this view, the butterfly's wings don't cause the tornado — they just cause our inability to say when the tornado will hit.

5

The Order Within

DESPITE ITS INNOCENT appearance, the tent map we examined in Chapter Four shows stunningly complex behavior. The tent map consists of two rigidly enforced rules applied one at a time. When state values are low, the tent map permits unfettered, exponential state growth. When state values are high, however, the map imposes severe penalties on further expansion. Thus it provides a crude mathematical model for what we observe all around us: no rapidly expanding process grows without limit. Explosions run out of fuel; industries exhaust their markets; populations eat up their nutrient resources. The tent map is instructive, but its simplistic structure misses some of the subtleties of real systems. In this chapter we turn our attention to another elementary nonlinear dynamical system, which is related to the tent map but is much nearer the mark of reality. As we will see, the chaos of this system (almost paradoxically) is *filled* with order. We will also see that some of the chaos of nature is similarly infused with regularity.

The Logistic Map

The logistic map has its origin in biological population dynamics. As we mentioned earlier, a population with unlimited nutrient supplies, perfect waste disposal, and no predators will, in principle, grow exponentially. But, of course, unlimited resources and perfect waste disposal are not achievable in practice. Even if a species has no predators, in a finite environment the waste products of the organisms will eventually poison the habitat and force individual members to compete for whatever nutrient remains accessible.

The tent map has in it a prohibition to unlimited growth, but the conditional character of the map—if x is less than 0.5 we have one dynamics, if x is greater than 0.5 we have another—is not typical of systems found in nature. In nature, limiting rules are usually continuously enforced. A better approximation to naturally occurring limited growth is obtained in the following model:

$$x_{n+1} = s \cdot x_n \cdot (1 - x_n) \tag{1}$$

A numerical example of iteration via equation (1) is $s = 1.5$, $x_0 = 0.3$. The first iterate, x_1, is equal to $1.5 \cdot (0.3) \cdot (1 - 0.3) = 0.315$; $x_2 = 1.5 \cdot (0.315) \cdot (1 - 0.315) = 0.3236625$. And so on.

We can once more think of x_n in equation (1) as the population of a single-cell colony relative to the environment's maximum possible population—its "carrying capacity." For this colony,

$$x_{n+1} = x_n + b' \cdot x_n - d \cdot x_n$$

where d (as in the tent map argument) is an intrinsic death rate, but b' is a birth rate depending on the amount of free space available:

$$b' = b \cdot (1 - x_n)$$

Here b is the intrinsic birth rate, the rate at which births would occur if unlimited free space were available. Measuring time in units of $1/d$ (along with a slight rearrangement) allows us to express the colony's population dynamics as equation (1). (The parameter s is the intrinsic birth rate of the population measured in rescaled time units.) The appearance of a quadratic term, x_n^2, in equation (1) results effectively from the pairwise competition between the cells for available free space. Equation (1) is called the logistic map. Like the tent map of the previous chapter, we do not mean to propose it as an accurate model of any real biological system. Our interest in the logistic map is in what it can reveal about the qualitative behavior of more detailed and elaborate models of reality.

Note that the logistic map and the tent map are closely related. When x_n is small, much less than 1, the term $(1 - x_n)$ in equation (1) is well approximated by 1. In this limit, x_{n+1} is approximately $s \cdot x_n$. That is, for low densities the population grows roughly exponentially — very much like what occurs in the tent map. On the other hand, as x_n approaches 1 (that is, nearing the carrying capacity), x_{n+1} becomes approximately $s \cdot (1 - x_n)$. The population in the $n + 1$st generation will therefore be small, reflecting the deleterious effects of competition for free space among the population's individuals. Recall that the tent dynamics for x near 1 is also $s \cdot (1 - x_n)$. Thus, in the two extreme limits of population size (for x near 0 and x near 1), the logistic map and the tent map are virtually identical. For intermediate population sizes, though, the two maps differ substantially. The consequences of these differences prove very interesting.

What does a graph of x_{n+1} versus x_n look like for the logistic map? A brief inspection of equation (1) shows that (a) x_{n+1} will be 0 when x_n is either 0 or 1 and that (b) if x_n is between 0 and 1, then both x_n and $(1 - x_n)$ will be positive, meaning that x_{n+1} will also be positive. This is because it is determined by the product of positive numbers (let's assume, for now at least, that s is positive — other cases are considered in Chapter Seven). Thus the dynamics curve associated with equation (1) will rise up from 0 when x_n is 0, achieve some positive values, and then fall back to 0 again when x_n is 1. A little numerical exploration suggests that the curve is symmetrical on both sides of $x_n = \frac{1}{2}$. (For example, when x_n is $\frac{1}{4}$, $1 - x_n$ is $\frac{3}{4}$; when x_n is $\frac{3}{4}$, $1 - x_n$ is $\frac{1}{4}$; and so on. Both choices of x_n produce the same product, the same x_{n+1} value.) Of course, how positive x_{n+1} gets will depend on the exact value of the parameter s. Example curves are shown in Figure 5.1.

The curve shown in the figure is called a parabola. As long as the top of the parabola doesn't exceed a height of 1, the course of the dynamics is completely contained in the square shown; that is, successive "populations" will be between 0 and 1. The

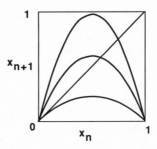

Figure 5.1 *Plots of x_{n+1} versus x_n different s-values.*

maximum of the parabola (as for the tent) occurs at $x_n = \frac{1}{2}$. How high is the parabola at its maximum height? That height is given by

$$x_{n+1} = s(\tfrac{1}{2})(1 - \tfrac{1}{2}) = \tfrac{1}{4}s$$

Remember that for the tent, the maximum height was s/2, implying that s had to be less than or equal to 2 in order to keep the dynamics bounded within the unit square. For the logistic map, s must be no greater than 4 for the same bounding condition. (The reason for these different values is that the tent permits unlimited growth to occur all the way up to $x = \frac{1}{2}$, whereas the logistic map folds into the dynamics limitation at all population sizes. The logistic map can consequently tolerate a much faster turnover rate—a larger s—before bursting through the x = 1 maximum.) A comparison of the two maps for the case s = 2 is shown in Figure 5.2.

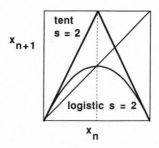

Figure 5.2 *The tent map and the logistic map for s = 2.*

Fixed Points of the Logistic Map

The dynamics of the tent map is fairly simple: when s is less than 1, 0 is an attractor; when s is greater than 1, there are strange attractors and the time series are chaotic. The logistic map yields much richer possibilities. To see that, we must first identify the fixed points of the logistic map. These occur where

$$x_f = s \cdot x_f \cdot (1 - x_f)$$

Clearly, $x_f = 0$ is one solution to this equation, which agrees with what we see in Figure 5.2: the dynamics parabola intersects the $45°$ line at the origin. (This result is the same as for the tent map.) There is a second intersection point with x_f between 0 and 1 if s is in the right range. To determine where, use the following argument: If x_f is not equal to 0, then we can divide both sides of the last equation by x_f and obtain

$$1 = s \cdot (1 - x_f)$$

or, after rearranging,

$$x_f = 1 - 1/s$$

In order for x_f to be between 0 and 1, s has to be at least as large as 1; otherwise $1/s$ will be greater than 1, and x_f will be negative. Thus what we have found is that if s is between 0 and 1, there will be only one nonnegative fixed point—at $x = 0$. When s is greater than 1, there will be two fixed points—at 0 and at $1 - 1/s$. The critical value $s = 1$ needed for the emergence of a second fixed point is the same as for the tent map (but where the second fixed point appears is different).

Characterizing the stability of the fixed points of the tent map is a simple matter. The slope of the tent at $x = 0$ is s, and at the second fixed point, when it exists, the slope is $-s$. When s is less than 1, $x = 0$ is stable (an attractor). When s is greater than 1, both fixed points are unstable (repellors). But what about the attracting or repelling nature of the logistic fixed points? What does it mean for something other than a straight line, such as the logistic parabola, to have a slope? The answer lies in the fact that the logistic parabola is a nice, *smooth* curve. As we argued in Chapter Three, very near any point on a smooth curve, the curve looks approximately like a straight line, and that straight line is the tangent line to the curve at the point in question. Determining the stability of the fixed points, then, requires knowing something about the slope of the tangent line to the actual dynamics curve at the fixed points.

How can we calculate the slope of the tangent line to a smooth curve at a given point on the curve? We can use a trick introduced by Newton over 300 years ago. Newton's trick is basically to use the idea that a smooth curve looks locally like its tangent line. Let's begin with a calculation of the slope of the tangent line at the fixed point $(x_n, x_{n+1}) = (0, 0)$ for the case $s = 3$. Take a nearby point $(x_n, x_{n+1}) = (0.1, 0.27)$ on the parabola. (If $x_n = 0.1$, then $x_{n+1} = (3)(0.1)(1 - 0.1) = 0.27$.) The slope of the tangent line at $(0, 0)$ can be approximated by the "rise over the run" of the line connecting the two parabola points chosen:

$$\text{slope} \approx \frac{0.27 - 0}{0.1 - 0}$$
$$= 2.7$$

But is this line really that close to the actual tangent line? See Figure 5.3.

We should try a succession of points ever closer to $(0, 0)$ with which to estimate the proper slope. For example, take the point

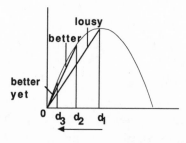

Figure 5.3   *Better and better approximations to the tangent line at (0, 0).*

(0.01, 0.0297). (Make sure you agree that this is a point on the parabola.) This time, the slope is approximated by

$$\text{slope} \approx \frac{0.0297 - 0}{0.01 - 0}$$
$$= 2.97$$

which is not the same value as before. Well, try an even closer point, (0.001, 0.002997):

$$\text{slope} \approx \frac{0.002997 - 0}{0.001 - 0}$$
$$= 2.997$$

Our calculations seem to be getting closer and closer to an answer of 3. That *seems* like an eminently plausible result (remember, we said that the logistic and tent maps were essentially identical at $x_n = 0$, and because the tent has a slope of s there, so, probably, should the logistic parabola). But how can we be sure? And how can we generalize our result at (0, 0) to some other point on the parabola (such as the other fixed point)?

Here's the general result we seek. The slope of the tangent line at $(x_n, x_{n+1}) = (x_n, s \cdot x_n \cdot [1 - x_n])$ can be found by taking a nearby point $(x_n + d, s \cdot [x_n + d] \cdot [1 - \{x_n + d\}])$, finding the slope of the line connecting the two points, and then letting d go to 0. In the latter limit, the slope of the approximating line and the actual tangent line will agree. A little algebra reveals that the tangent line's slope is

$$\text{slope} \approx \frac{s \cdot [x_n + d] \cdot [1 - \{x_n + d\}] - s \cdot x_n \cdot [1 - x_n]}{x_n + d - x_n}$$
$$= -2sx_n + s - sd$$

In the limit as d vanishes, this becomes

$$\text{slope} = -2sx_n + s$$

This equation gives the slope of the tangent line attached to the logistic parabola above *any* value x_n. You can see that if $x_n = 0$ in this general result, the desired slope is s (that is, 3 in our numerical example). At $x_n = \frac{1}{2}$, the slope of the tangent line to the logistic parabola is 0 — the tangent line is horizontal there. (Do you see why from Figure 5.1? Draw tangent lines to the three example curves. Note also that $x_n = \frac{1}{2}$ is where the curve turns around; it stops going up and starts going down. This is a general characteristic: wherever a curve turns around — wherever it has a local maximum or minimum — the slope of its tangent line at the turnaround point is zero.) At $x_n = 1$, the slope is $-s$ (just like the tent). Thus as we travel along the x_n-axis, the slope of the parabola varies from $+s$ (on the left) to $-s$ (on the right), taking on all values in between.

("Slope of the tangent line" is a bit cumbersome. Traditionally this phrase is replaced by the not-so-informative term *derivative*. Before, we said that a curve was smooth if it had a well-defined tangent line at every point. Alternatively, we can now say that a curve is smooth if it has a derivative everywhere. The derivative has a very useful meaning: it is the rate at which the curve it derives from is changing at any specified point on the curve. The vanishing of a derivative is a signal that the curve it derives from has a local extremum at the vanishing point. Derivatives are therefore powerful tools in answering such questions as "How long must we wait for such-and-such to be a minimum?" and "How much input will maximize the output?" Derivatives are part of the apparatus of calculus. The roots of calculus are embedded in the study of motion — rate of change of position — but its applicability and appropriateness to the study of change in general have long been recognized. The importance of calculus in the development of science and technology cannot be over-stated, nor can its central place in Western culture be denied.)

The nonzero fixed point, when it exists, occurs at $x_f = 1 - 1/s$; the slope of the tangent line to the logistic curve at the second fixed point is $-2s(1 - 1/s) + s = -2s + 2 + s = 2 - s$. It is *not* $-s$, as in the tent map. This seemingly small difference between the two maps results in a very strong difference between their two dynamical outputs. We can now answer the question about the stability of the fixed points of the logistic map. As long as s is less than 1, $x_n = 0$ is the only fixed point, and because the slope of the tangent line there is s, it is stable. But when s becomes greater than 1, the origin becomes a repelling fixed point, and a second fixed point appears at $1 - 1/s$. At first, when s is only a little bigger than 1, the slope, $2 - s$, of the tangent line at the second fixed point is positive and less than 1. When s = 2, the second fixed point is at $\frac{1}{2}$ (the top of the parabola), and the slope of the tangent line is 0 (the slope of the tangent line to the parabola at $x = \frac{1}{2}$ is always 0). When s becomes greater than 2, $2 - s$, the slope at $1 - 1/s$, goes negative. At s = 3, the slope becomes −1. This is an important value: for s between 1 and 3, the second fixed point is stable (attracting). For s greater than 3, though, the second fixed point becomes unstable, and the dynamics begins to meander around on the parabola.

Surprises in the Dynamics of the Logistic Map

We now know that special things (bifurcations) happen in the logistic map at s = 1 and at s = 3. (Remember that the tent map has only a single bifurcation, at s = 1.) Our investigation of the tent map led us to the conclusion that the dynamics is characterized by an attractor for every control parameter value (s). These attractors can be visualized by calculating a succession of x-values and plotting the latest ones (to get rid of idiosyncratic start-up transients). To see what interesting dynamical behavior the logistic map supports, we cut directly to the chase: Figure 5.4 is the bifurcation diagram of the logistic map.

A comparison of Figure 5.4 with Figure 4.40, the bifurcation diagram of the tent map, highlights the many more interesting features of the former. There is a first bifurcation at s = 1, corresponding to the onset of instability of $x_n = 0$. A second bifurcation at s = 3 corresponds to the onset of instability of the nonzero fixed point. For s just a little greater than 3, two very narrow prongs seem to emerge, suggesting a 2-cycle behavior. A similar two-pronged fork was seen in the tent bifurcation diagram, but there the seeming periodic behavior proved illusory: though x_n roughly repeated every other iteration, in exact detail it did not. Is this the same behavior we are witnessing in the

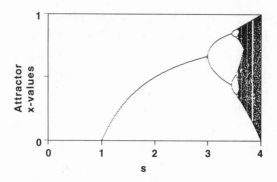

Figure 5.4   *The bifurcation diagram of the logistic map.*

logistic bifurcation diagram? Careful scrutiny of the time series for s = 3.1 (after allowing 200 iterations to "wipe away" any start-up peculiarities) reveals that x_n repeatedly hops back and forth between the values 0.7645665199585943 . . . and 0.5580141252026960 . . . .This is a real 2-cycle! Similarly, just beyond about s = 3.5, the bifurcation diagram appears to be four sharp prongs. Again, careful scrutiny of the time series for s = 3.5 (after enough start-removing iterations) yields successive x_n values 0.8749972636024641 . . . , 0.3828196830173241 . . . , 0.8269407065914386 . . . , and 0.5008842103072180 . . . , repeated continually. The structure suggested in the bifurcation diagram is indeed a 4-cycle.

What dynamical behavior does the logistic map demonstrate after the 4-cycle? To see, we magnify the region between s = 3.4 and s = 3.6, in Figure 5.5. At the left of the figure are the

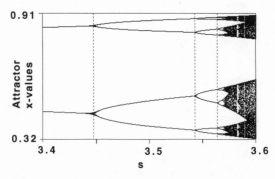

Figure 5.5   *The region between s = 3.4 and s = 3.6.*

2-cycle "tracks." At about s = 3.45, the 2-cycle bifurcates into a 4-cycle. Near s = 3.55, the 4-cycle bifurcates into an 8-cycle, and shortly thereafter (in parameter value, not time), the 8-cycle becomes a 16-cycle. In each successive bifurcation the cycle that emerges has a repeat time *twice* as long as that of the previous cycle. Closer to s = 3.6, the sharp prongs that are symptomatic of cyclic behavior thicken, much like the onset of confined chaos that we observed in the tent map. A careful study of time series does indicate that nonrepeating characterizes the logistic dynamics for s near ·3.6. There is a nice phrase describing the behavior shown in Figure 5.5: "Chaos appears at the end of a cascade of period-doubling bifurcations." In this scenario chaos is the natural termination, as the control parameter is increased, of a sequence of cycles, each with period twice as long as the previous. Eventually, the period is "so long" that the time series never repeats.

Well, that's pretty interesting, but the bifurcation diagram of the logistic map, Figure 5.4, has still more structure (than its cousin for the tent map did). Figure 5.6 is a magnification of the region of the bifurcation diagram from s = 3.5 to s = 4. The thickened, presumably chaotic attractors in this region are interspersed with open gaps. Closer inspection of these gaps shows that they are actually connected by narrow prongs, suggesting periodic attractors within. Chaos is interspersed with "windows of periodic order." The most prominent of the periodic windows occurs near s = 3.8. Figure 5.7 magnifies this window. Near

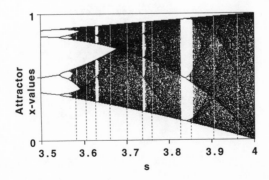

Figure 5.6   *The region of the logistic map bifurcation diagram from s = 3.5 to s = 4. The leading edges of a few windows of stable periodicity are noted.*

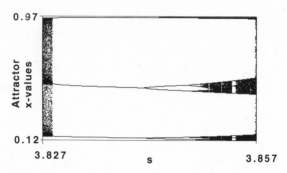

Figure 5.7   *The 3-cycle window.*

$s = 3.83$, chaos suddenly terminates and a period 3-cycle emerges. There is curious structure within this window, which on the scale of Figure 5.7 is difficult to decipher. Seemingly, as s is increased slightly, the 3-cycle bifurcates into a 6-cycle; at higher s, chaos reappears. Does the 3-cycle window contain a period-doubling cascade?

We look at only the middle prong of the 3-cycle, magnifying that in Figure 5.8. In order to produce the enlarged copy of the tiny piece of the bifurcation diagram shown to the right in the figure, the enlargement was flipped top-to-bottom and magnified by somewhat different amounts in the vertical and horizontal directions. The resulting image is remarkably similar to the whole bifurcation diagram (including a main sequence of period doublings terminating in chaos and a fat "3-cycle" window; note that because this piece of the bifurcation diagram is one-third of

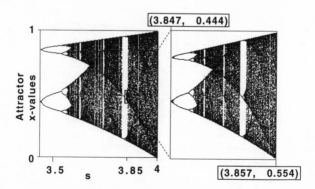

Figure 5.8   *Comparison of the bifurcation diagram of the logistic map for s = 3.5 to s = 4, with a small piece of the diagram located near s = 3.85.*

all that happens in the large 3-cycle window, the "3-cycle" is really a 9-cycle).

Each of the windows marked in Figure 5.6 is like the 3-cycle window of Figure 5.7, except that emerging from the preceding band of chaotic behavior are cycles of different periodicity. The relatively large window starting near s = 3.73 is a 5-cycle window. That near s = 3.63 is a 6-cycle window. The much narrower window near s = 3.7 is a 7-cycle window. Near s = 3.91, there is another 5-cycle, near 3.96 another 4-cycle, and so on. There are windows of all orders embedded in the bifurcation diagram; the periodicities of these windows are scrambled in a complicated way. Among the general rules of periodic windows is "The higher the periodicity of the window and the closer it is to s = 4, the narrower it is." A period-100 window is so narrow that it is extremely difficult to find by conducting a numerical hunt through successive s-values. (If you had no idea where to start your search, you would almost certainly either skip past the window because you would be incrementing s by too large a value in each step, or take so many steps that you would never finish.) Amazingly, the cyclic attractors that emerge from chaos in each of these windows bifurcate (with increasing s) into attractors with twice the original period, then four times, then eight times, then. . . . Each prong of the original attractor completely recapitulates the overall story of the whole bifurcation diagram; each prong, suitably enlarged, is a shrunken copy of the whole diagram. In this sense, the bifurcation diagram of the logistic map — its family portrait of attractors, periodic and chaotic — is fractal-like: it is filled with copies of itself on many different length scales.

Why are the periodic cycles of the logistic map so different from those of the tent map? The tent map contains *no* stable cycles, whereas the logistic map contains stable cycles of *all* periods. The answer lies in a study of the higher-order return maps associated with the logistic map. As in the case of the tent map, a plot of x_{n+2} versus x_n tells us about the nature of 2-cycles. Figure 5.9 shows second-return maps for the logistic

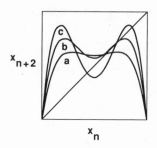

Figure 5.9   *Second-return maps for the logistic map for (a) s = 2.5, (b) s = 3.0, and (c) s = 3.5.*

map for three values of the control parameter. When s is less than 3, the plot of x_{n+2} versus x_n is a curve that cuts the 45° line at the fixed points of x_{n+1} versus x_n. As s is made greater and greater, the second-return map becomes a two-humped, smooth curve with its middle valley getting ever deeper. This structure eventually cuts through the 45° line in four places: at the two fixed points of the first-return map and at two other x_n-values that constitute a 2-cycle. Because the logistic map and all of its higher-order return maps are smoothly curved (unlike the analogous jagged straight-line-segment structures of the tent map), their tangent lines can have a wide range of slopes at the cut points. Indeed, for s between 3 and about 3.45, the slopes of the second-return map tangent lines at the 2-cycle values are less than 1 in magnitude. The 2-cycle is therefore stable in this s-range. See Figure 5.10. (Note the nonintuitive result that the tangent lines at the two 2-cycle values have the same slope.)

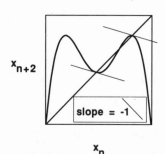

Figure 5.10   *Slopes of the logistic second-return map at the 2-cycle values. The case shown is for s = 3.3. A line with slope equal to −1 is drawn for reference.*

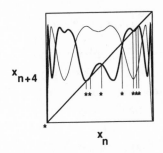

Figure 5.11   *The logistic fourth-return (bold) and third-return (light) maps for s = 3.5.*

When s is around 3.5, the 2-cycle becomes unstable. At exactly the same value of s at which this happens, a stable 4-cycle appears. Figure 5.11 shows a plot of x_{n+4} versus x_n for s = 3.5. The fourth-return curve nicks the 45° line in eight places: two where the fixed points of the first-return map are, two where the (now unstable) 2-cycle values are (because a 2-cycle is also a 4-cycle), and four other places determining the values of x_n that constitute a 4-cycle. The latter, near s = 3.5, is stable. The figure also shows the curve of the third-return map in light in the background. Note that at s = 3.5, this curve intersects the 45° line only twice — at the fixed points of the first-return map. No 3-cycle is possible in the logistic map for s = 3.5. It isn't until s is about 3.83 that a 3-cycle appears (that is, at the onset of the 3-cycle window).

The 4-cycle associated with Figure 5.11 becomes unstable near s = 3.55. A search through the higher-order return maps reveals that at the point of instability of the 4-cycle, a stable 8-cycle becomes possible. This sequence of appearance of ever higher-order cycles produces the period-doubling cascade described previously. In general, the higher the order of the cycle, the more writhings its associated return-map curve has. Very slight changes in control parameter produce quite large changes in the depth and steepness of these writhings. Stable cycles come and go in a complicated pattern as s is increased, and the intervals of s over which they are stable get smaller and smaller with increasing order.

Universality

In the 1970s, Mitchell Feigenbaum made a profound discovery about the logistic map: the period-doubling route to chaos has *universal* characteristics [87]. That is, behavior similar to the logistic map (in a precise sense that we will make clear below) can be seen in very dissimilar systems. The story of universality begins with the bifurcation diagram. Look at Figure 5.12. Let's denote the values of s where period doublings occur in the primary cascade by $s_1, s_2, s_3, s_4,$. . . . The value s_1 is 3; s_2 is close to 3.45; s_3 is close to 3.55, and so on. Note that these s-values get closer and closer the deeper you go into the cascade. Feigenbaum calculated the following ratio for a number of bifurcations:

$$\frac{s_k - s_{k-1}}{s_{k+1} - s_k}$$

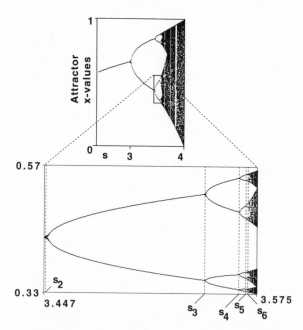

Figure 5.12   *The bifurcation diagram of the logistic map (above) and an enlarged portion of the primary period doubling cascade.*

For example, using the approximate values given above, we get

$$\frac{s_2 - s_1}{s_3 - s_2} \approx \frac{3.45 - 3.00}{3.55 - 3.45} = \frac{0.45}{0.10} = 4.5$$

What Feigenbaum observed was that as k becomes larger and larger (that is, as you go deeper and deeper into the cascade), these ratios approach a common value, 4.669201609. . . . This so-called Feigenbaum number measures the spacing—in control parameter values—of successive bifurcations. The spacings get smaller and smaller by amounts approximating the fixed ratio 4.6692. . . .

At first it may seem unlikely that the Feigenbaum number is as fundamental as pi ($\pi = 3.141592653589$ . . . , which shows up wherever periodicity is found, such as in the geometry of the circle or the undulation of the sine curve) or the exponential constant, e (2.718281828459 . . . , which characterizes exponential growth and decay). A first hint that this is so, however, lies in the fact that the period-doubling cascades found in all the other periodic windows of the logistic map bifurcation diagram have the same spacing structure as the main cascade, if you look deep enough into the cascade. The second hint is contained in the following remarkable fact: If you replace the parabola of the logistic map with other "one-hump, round" shapes, then (a) you get period doubling preceding chaos as you increase the "height" of the round function, and (b) the same 4.669 . . . ratio characterizes the spacings of the successive doublings. In other words, the onset of chaos through a cascade of period-doubling bifurcations has properties that are independent of the details of the dynamical rules that produce the behavior.

A second example of a "one-hump, round" dynamical system is the sine map,

$$x_{n+1} = \left(\frac{s}{\pi}\right) \cdot \sin(\pi \cdot x_n)$$

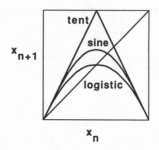

Figure 5.13   *A comparison of the tent, logistic, and sine maps, all for s = 2.*

This definition of the sine map vanishes at $x_n = 0$ and $x_n = 1$, like the tent and logistic maps, and its tangent line has the same slope as both of these maps at $x_n = 0$ and $x_n = 1$. The quantity s, then, has a similar meaning in all three maps. The sine map is shown in Figure 5.13, along with the tent map and the logistic map, all for the same value of s (s = 2). The relative heights of the three maps remain as shown in the figure for any s. The tent map has no bounded attractors (that is, x_n blows up rather than staying between 0 and 1) for s greater than 2, and the logistic map has none for s greater than 4. The sine map, being intermediate in height between the other two, loses its bounded attractors at an intermediate value of s: $s = \pi$.

Figure 5.14 compares the bifurcation diagrams for the tent map (T), the sine map (S), and the logistic map (L). Note how similar the sine and logistic bifurcation diagrams are to each other and how dissimilar they are to the tent diagram. The tent has only two slopes (+s and −s), whereas the round maps have a

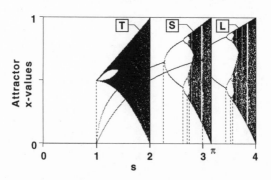

Figure 5.14   *Three bifurcation diagrams.*

continuum of slopes (between +s and −s); the dynamics of the tent map is less rich as a consequence. Despite detailed quantitative differences between the logistic and sine maps, the period-doubling cascades in their two diagrams are characterized by identical values of the Feigenbaum number, 4.669. . . .

It's all well and good to talk about the interesting behavior of mathematical games like the logistic map, but one also wonders whether these games have any practical use. When we first started discussing the logistic map, we noted that it had its roots in population biology. We also noted, though, that the logistic map is highly simplified; we would hardly expect it to describe a real population very well. Our hope in examining its properties was to be guided in thinking about the qualitative behavior of real dynamical systems. That's why the following empirical observations are so surprising.

A simple electronic circuit. An extremely simple electronic circuit with extraordinarily complicated behavior can be constructed from a coil of wire (an inductor), a diode (a semiconductor device that allows current to flow through it easily in one direction and poorly in the other), and a generator of alternating current. To monitor what is happening in the circuit, we hook up oscilloscopes (TV sets that read electrical currents) across the generator and the diode. The circuit is represented schematically in Figure 5.15.

Given "its druthers," the generator will attempt to cause there to flow in the circuit a periodically varying current like the one

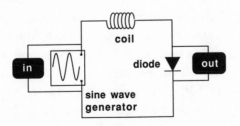

Figure 5.15   *A diode, coil, and generator circuit. The dark blobs represent oscilloscopes to keep track of what is happening.*

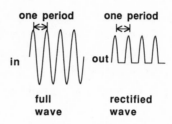

Figure 5.16   *A full sine wave is put into the circuit by the generator (in) and rectified into half a wave by the diode (out).*

shown on the left of Figure 5.16. The behavior of this circuit is a little different from the simple maps we have been discussing in that the current is not updated at discrete intervals but, rather, is *continuously* changing. A full treatment of continuous dynamics requires subtle mathematical tools such as calculus and differential equations. On the other hand, continuous dynamics can be converted into discrete dynamics by a trick called a Poincaré map. What we do is sample the output of the system—the current, in this case—at appropriate intervals. We are driving the circuit with a periodic signal, so a convenient sampling rate is the frequency of the driving signal. Thus if the signal generator has a frequency of 1 million cycles per second, say, then we sample the current every millionth of a second. (That sounds hard, but it can be done with the help of a computer and some fancy electronics.) This produces a discrete list: $current_1$, $current_2$, $current_3$, . . . , where the subscripts mean "at time 1," "at time 2," "at time 3,. . . . " The Poincaré map of these data is a plot of $current_{n+1}$ versus $current_n$, a first-return map with the "generations" determined by the sampling rate. Because the continuous dynamics is deterministic, we expect that there will also be a deterministic relationship between successive currents sampled in this way and that the plot will help us figure this out.

In any event, let's take the period of this input to be 1. If the diode were simply a resistor, the oscilloscope reading of the current through it would be very similar to the input signal; it would be another full sine wave with exactly the same period. Depending on the operating conditions, a diode can allow the flow of current in the "forward" direction but can prohibit flow in the "backward" direction. Under these conditions, an oscilloscope monitoring the current through the diode shows half of a sine wave — like the one on the righthand side of Figure 5.16. Though the "negative" half of the current is chopped off by the diode's action, the signal still varies periodically in time with the same period as the input signal.

A diode is not merely a resistor. Up to a point, the more current you push through it in the easy direction, the easier it gets to push even more current. Diodes are inherently nonlinear. The circuit of Figure 5.15 is a dynamical system whose control parameter (the thing analogous to the height of the parabola in the logistic map) is the driving force for pushing current through the circuit — the voltage across the poles of the generator. Figures 5.17a–e are photographs of a display of an oscilloscope connected across a diode in an actual circuit like that of Figure 5.15. When the driving voltage is low, the display of the oscilloscope across the diode looks like screen a: it shows a periodic signal with the same period as the input sine wave. When the driving voltage is increased sufficiently, the signal across the diode undergoes a significant change. Every other wiggle changes

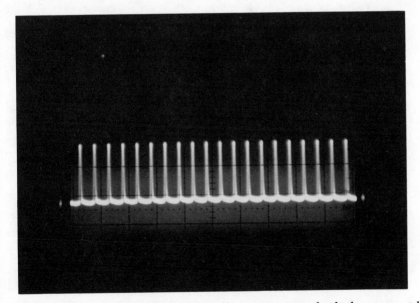

Figure 5.17a   *At low applied voltages, the output across the diode repeats with the same frequency as the generator input.*

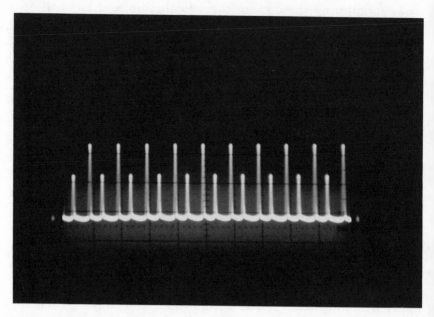

Figure 5.17b *At a higher applied voltage, the time for the diode output to repeat is twice as long as the input period.*

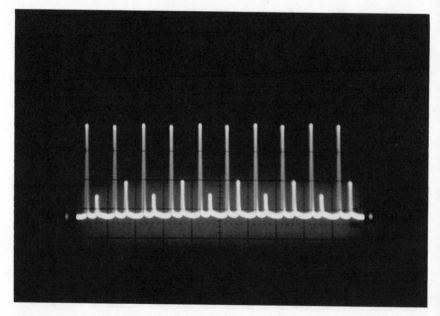

Figure 5.17c *At a still higher applied voltage, the output repeat time is four times the input period.*

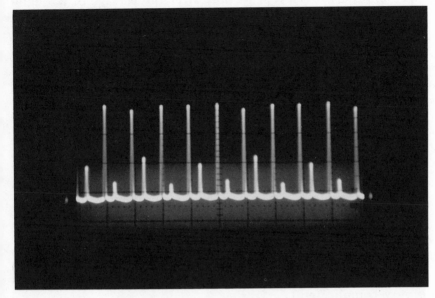

Figure 5.17d *A slightly higher applied voltage still produces an output period eight times as long as the input.*

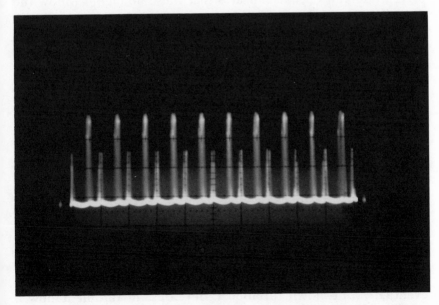

Figure 5.17e *Finally, a small increase in applied voltage produces chaos.*

size, and the pattern repeats only after the passage of *two* input periods (see screen b). This is a real-world period-doubling bifurcation. Further increases in the driving voltage lead to what is seen on screens c and d — signals of periods 4 and 8, respectively. Though in principle an infinite cascade of such bifurcations might be expected, relatively few more can be observed. (Remember that the doublings occur ever more closely the deeper we go into the cascade; after a while, our control knobs can no longer tweak the circuit with sufficient precision for us to discern the doublings.) Ultimately, the diode oscilloscope display can no longer be stabilized, much as on screen e. The peaks seem to wiggle erratically. The output of the diode never exactly repeats. This is electronic chaos. (Careful increased adjustment of the driving voltage beyond the onset of chaos leads to the discovery of periodic windows in which wave forms that repeat with periods 5, 3, and so on, appear. The observable behavior of such circuits is very complicated.)

Thus a very simple alternating-current circuit shows a cascade of period doublings leading to chaos. Quite unexpectedly — because the logistic map has nothing whatever to do with this circuit — a measurement of the successive voltages at which period doublings occur indicates that these doublings follow the Feigenbaum sequence to within experimental uncertainty [88]. Back in Chapter One we mentioned that noise exists in all kinds of electrical devices. The electrical chaos discussed here is *not* related to that noisy behavior. Current fluctuations in a resistor are minuscule, and they occur even when a constant voltage is applied. Chaotic current fluctuations in diode circuits are enormous by comparison, and they appear only when the applied voltage is rapidly changing.

Turbulence in Fluid Flow. The slow flow of a fluid confined in a pipe is said to be laminar. A fleck of dust inserted into laminar flow travels along a straight line. As the flow speed is increased, undulations in the flow can develop far from the walls of the pipe. At first these undulations are regular waves with well-defined periods. As the flow speed is increased even further, the periodic waves become successively more irregular, and eventually full-blown turbulence develops. In turbulent flow a fleck of dust executes crazily erratic motions. The passage from laminar flow to turbulence can be seen in different fluid experiments. One, reminiscent of Lorenz's study of atmospheric convection, involves heating a confined liquid from below and establishing convective churning. With such a set-up, Albert Libchaber, in France, observed a period-doubling cascade prior to turbulent behavior with the Feigenbaum number characterizing the cascade [89].

Huge numbers of electrons and huge numbers of molecules constitute the flows described in the preceding examples. In the orderly regimes, these many particles act collectively, moving together in lock step. An adequate description of the flow can be attained with a small number of variables, such as the average local flow velocity. A natural guess about the transition to chaos is that this collective state falls apart and the many individual pieces begin executing independent motions. But that is not the case: even in chaotic, turbulent flow, the particles of the stream move collectively. The parameters that adequately describe the orderly state are sufficient to describe the chaotic one as well. The existence of the Feigenbaum route to chaos in these systems demonstrates that their apparent erratic behavior is built on a firm, orderly, deterministic foundation.

Heart Beats. The heart is an organ that exhibits rhythmic behavior (though, as we've said before, not exactly periodic behavior). In the early 1980s, Leon Glass and colleagues used chick heart muscle cells in some experiments that are very suggestive and are perhaps of fundamental clinical importance as well. The cells are separated in a Petri dish and immersed in nutrient. One of the cells is electrically stimulated externally. At low stimulation levels, the chick cells undergo simultaneously synchronized contractions (presumably through chemical interactions with neighboring cells). At higher stimulation levels, the contractions show period doublings and suggest the Feigenbaum structure. Eventually, chaotic contractions are evidenced throughout the dish—a kind of model fibrillation [90].

Many other examples of the period-doubling route to chaos have been observed. Though it is hard to see, the light from a laser actually flickers in intensity. Periodic flickering can be made chaotic by adjusting the length of the laser cavity; these periodic flickerings can be made to undergo a period-doubling succession as chaos is approached [91]. Similarly, there are a few chemical reactions (the Belusov–Zhabotinski reaction being the most famous) that oscillate in time. Careful control of the flow rates of the reactants can lead to chaotic output following period doubling [92]. The time interval between the release of drops from a slightly open faucet has been observed to include a period-doubling route to chaos [93]. Period doubling leading to chaos has even been reported in woodwind instruments [94]. We mention finally a speculative work of Alvin Saperstein, who has examined the ratio of armaments in the Soviet Union to those in Nazi Germany in the years preceding World War II. He believes that there is evidence for period doubling in this ratio and further asserts that this ratio may well have achieved fully chaotic behavior by the time the war started. Although this study is based on limited data, it is certainly a titillating premise that systems as complex as societies can in some sense be viewed as nonlinear dynamical systems behaving in ways qualitatively similar to the logistic map! [95]

None of the examples cited here is expected to be described in any quantitative sense by the logistic map. Nonetheless, they all have in common the period-doubling route to chaos. Period doubling is not the only scenario by which deterministic chaos can appear, but when it does, it seems to follow a universal sequence that is epitomized by the Feigenbaum number. In this sense, the Feigenbaum number is one of the central numerical constants of the universe. Just as π describes the geometry of periodicity, and just as e describes the geometry of exponential growth, the Feigenbaum number describes the geometry of one extremely important aspect of chaos.

6

Hide and Seek

WE SAW IN the previous two chapters that chaos can be present in the output of even extremely simple dynamical systems. The seemingly erratic changes associated with chaotic behavior test our ability to predict. Even when we know the dynamical rules perfectly, small, ever-present measurement uncertainties rapidly amplify, making exact, long-term forecasting problematic. But deterministic chaos is very different from random occurrence, where even short-term prediction is impossible. Deterministic chaos, as we argued in Chapters Four and Five, is filled with order. The challenge is to detect that order and, once it is detected, to harness it for prediction and control.

Prisms, Pianos, and Spectral Analysis

Figures 6.1a–f portray some of the many faces of noisy time series. Though none of the time series shown ever repeats, visual inspection suggests a qualitative difference in the irregularity from figure to figure. The apparent patterns in the noise seem to imply the existence of rules or mechanisms, perhaps different for each time series. The eye is an excellent pattern recognition device, but it is not very quantitative. Questions such as whether the differences between Figures 6.1a and 6.1b are greater or less than the differences between Figures 6.1a and 6.1f, for example, are extremely difficult to answer by eye. We need some analytic tools to help us make such judgments.

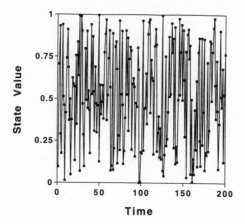

Figure 6.1a   *A uniformly random time series.*

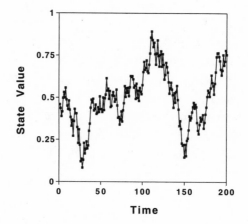

Figure 6.1b   *A random walk.*

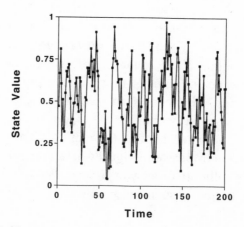

Figure 6.1c   *1/f noise.*

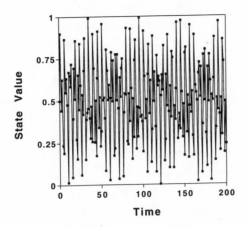

Figure 6.1d   *Quasiperiodicity.*

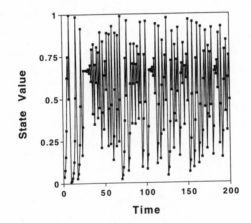

Figure 6.1e   *Time series from the tent map.*

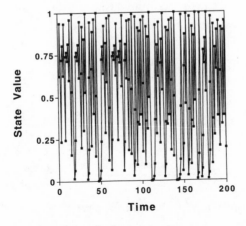

Figure 6.1f   *Time series from the logistic map.*

The first tool, one of the most widely used in the study of noise, has the scary name *spectral analysis*. You may not realize it, but you've probably seen and/or heard spectral analysis before. One familiar example occurs when white light passes through a prism or a water droplet (Figure 6.2). The light seems to spread out into "colors of the spectrum." Another, less familiar example requires a piano. If you clap your hands hard while standing next to a piano and then listen carefully, you will hear a number of the piano strings "ringing."

Both light and sound are oscillatory phenomena. Light consists of ripples in the electric and magnetic fields that bathe our retinas. Sound is ripples in the pressure the air exerts on our ear drums. A pure *color* is a light oscillation of a single frequency. (We can see electromagnetic ripples ranging in frequency from about 4×10^{14} Hz to about 7×10^{14} Hz. The unit 1 Hz is shorthand for "one Hertz," or "one cycle per second.") White light is a mixture of colors (so many colors that our eye gets confused and calls the mixture white). The prism or droplet can unmix white light because each of the component colors travels at a slightly different speed through glass or water. A "dispersion" of the white light into its component colors occurs when the light is bent by the prism or droplet; slower blue light is bent more than faster red light.

Similarly, a pure *tone* is a sound oscillation of a single frequency. (Young ears can hear pressure ripples ranging in frequency from about 20 Hz to about 20,000 Hz — the upper frequency that old ears can detect is much lower.) A clap is a mixture of pure tones (so many that our ear gets confused). A piano can unmix the clap because each of its strings vibrates at a different "fundamental" frequency. When a string is jiggled by a vibration corresponding to its fundamental frequency, it responds

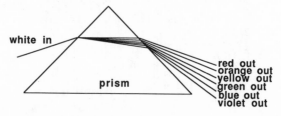

Figure 6.2   *Dispersion of white light by a prism.*

by also vibrating at that frequency. (This "sympathetic" vibration is called resonance.) A clap, made of many sound frequencies, can jiggle many of the strings in the piano. Of course, a piano isn't a great unmixer because it has only 88 strings (each with its own fundamental frequency). A device with an infinite number of strings would do a much better job!

Spectral analysis of light consists of identifying all the frequencies present and measuring the brightness of each associated color. Spectral analysis of sound consists of identifying all the frequencies present and measuring the loudness of each associated tone.

The relevance of prisms and pianos to detecting deterministic causes in noisy time series becomes clearer when we (1) recognize that a pure color or a pure tone, like the purely periodic oscillations of a simple pendulum, can be modeled by a sine function, and (2) believe the astonishing mathematical fact that almost any functional form—including jagged, erratic noise—can be represented as a sum of sines. Let's begin by examining the first part of this statement.

Figure 6.3 shows 10 complete oscillations of a sine function. The sine shown in the figure exactly repeats every 1 unit of time. For a pendulum, a typical repeat time might be about 1 sec. For a pure musical tone, a typical repeat time is a few milliseconds—a few times 10^{-3} sec. For visible light, the repeat time is a few femtoseconds—a few times 10^{-15} sec. The *frequency* of any sine

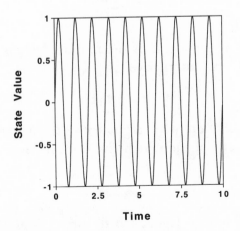

Figure 6.3 *A sine function.*

function is the number of repetitions of the function per unit
time, a number identical to the reciprocal of the repeat time.
Thus, for a pendulum, a typical frequency might be about 1 per
second — 1 Hz; for a tone, on the order of 10^3 Hz; for light, on
the order of 10^{15} Hz.

Here's a typical sine function: sin(A). What we call the argument, A, of the sine function — the stuff inside the parentheses
— is an angle. So how does time enter the picture? That is, in
what sense can we say that a sine repeats after a certain amount
of time? If we make a plot of the sine as a function of its
argument, A, we see that it wiggles exactly like the graph in
Figure 6.3, provided that A increases at the same rate as time —
that is, provided that when the time changes from t_1 to t_2, A
changes from A_1 to A_2, and

$$A_2 - A_1 = B \cdot (t_2 - t_1)$$

What should we take B to be? Well, when the time changes by
one full repeat time, the sine is supposed to repeat exactly,
which means that A changes by $360°$ or, equivalently, by 2π
radians. Letting $t_2 - t_1$ be T — the repeat time — and letting
$A_2 - A_1$ be 2π — the repeat angle measured in radians (angular
measure is less messy in radians) — we find that

$$B = \frac{2\pi}{T}$$

Note that the frequency, f, of the motion described by the sine
function is $1/T$, so, alternatively,

$$B = 2\pi \cdot f$$

Thus a sine function with argument $2\pi \cdot f \cdot t$ can represent a
simple oscillatory phenomenon that repeats every $1/f$ units of
time.

The vertical axis in Figure 6.3 denotes values of the relevant
physical parameter. The maximum possible value that the param-

eter can take on is called the *amplitude* of the oscillation. In the figure, state values are measured in units of the appropriate amplitude. That is, 1 on the vertical axis represents 1 unit of the appropriate amplitude. For the pendulum, the relevant state value is the displacement of the pendulum bob relative to when it is hanging vertically downward. If it were describing a pendulum, then, the figure would be a record in time of the displacement of the pendulum from equilibrium. The amplitude of the motion, a distance, might be measured in centimeters, say. "Plus" denotes displacement in one direction (such as to the right), and "minus" denotes displacement in the other direction (respectively, the left). The energy of the pendulum is proportional to the square of its amplitude; to double a pendulum's amplitude, you have to increase its energy by a factor of 4.

On the other hand, if the figure were describing light, the relevant parameter might be the electric field strength at a point on your retina. Typical electric field strength amplitudes in light range from less than 1 volt per meter in dim light to over 10^4 volts per meter in a bright laser beam. An electric field points in a direction, so "plus" might designate one direction and "minus" the opposite direction. Similar to the case of the pendulum, the brightness of light (its energy) is proportional to the square of the amplitude of its electric field strength. For a tone, the relevant parameter might be the difference in pressure of the air, at your ear drum, from that of "quiet" air (about 1 atmosphere). Typical sound amplitudes range from 10^{-9} atmospheres in soft sounds up to about 10^{-4} atmospheres at a loud rock concert or near a jet taking off. In this case, "plus" denotes greater than quiet air pressure; "minus" denotes less. Loudness (the sound energy) is proportional to the square of the amplitude of the pressure difference.

Although you may be aware of the kinship between oscillatory phenomena and the geometry of sines, the second part of the statement we made on page 191—almost any functional form can be represented as a sum of sines—may be less well known to you. Suppose we wish to construct a most un-sinelike functional form, such as the triangular ramp shown in Figure 6.4. The first five sines necessary to construct this ramp are shown in Figure 6.5. Each successive sine function has a higher frequency and a

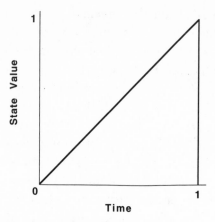

Figure 6.4   *A triangular ramp function (bold).*

smaller amplitude. When added together, these five sines produce the wiggly form shown in Figure 6.6.

Because the ramp is not smoothly rounded, many more terms
—at higher and higher frequencies—are required to make a
very good fit. Still, you can see that the sum of smoothly curved
sine shapes can actually approximate the sharply broken triangular form. How the sines are selected to synthesize the desired
shape (that is, what the proper frequencies, amplitudes, and
relative placements are) is beyond the scope of our discussion.
What we want you to remember is that any function whose graph

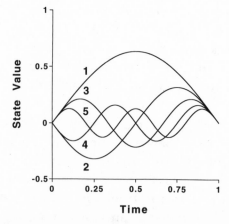

Figure 6.5   *The first five sine functions needed to synthesize the triangular
ramp.*

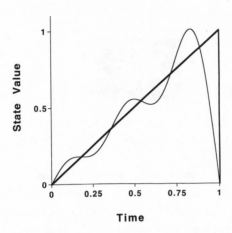

Figure 6.6   *Five-term synthesis of the triangular ramp.*

we can draw—including ones that are broken and jagged—is composed of a sum of periodically wiggling functions. Because this idea was discovered by the French mathematician Jean-Baptiste-Joseph Fourier in the early 1800s, this sum is called the Fourier synthesis of the function in question. (A musical synthesizer mixes pure tones—pure sines—together to create musical sounds reminiscent of violins, harpsichords, flutes, and so forth. Musical instruments do not produce pure tones; always present in their sounds are mixtures of tones that are said to give the instruments their characteristic "colors.") The inverse process—pulling a function apart into periodic components—is called Fourier analysis.

When we pass white light through a prism, and when we clap near a (many-stringed) piano, we are decomposing those signals into their constituent, periodically wiggling sine functions. Spectral analysis *is* Fourier analysis. We can imagine a kind of generalized prism, a "black box" filled with the right physical apparatus, into which we can feed an arbitrary signal and out of which comes a read-out of the signal's sine function components. Such a generic device is called a spectrum analyzer. A schematic diagram of how a spectrum analyzer might work is shown in Figure 6.7. (Compare this diagram with Figure 6.2.) In Figure 6.7, the input is some wriggly shape. The output consists of three sines with the amplitudes and frequencies shown: the wriggly shape is the sum of the three sines.

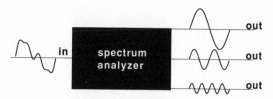

Figure 6.7   *The input to a spectrum analyzer is a complicated form, and the output is a series of periodic functions.*

In practice, because data collection in the real world requires individual measuring and recording events, the input to be analyzed is usually in the form of a set of discrete values—a table or a list—not a continuous wave form like that shown in Figure 6.7. For example, discrete inputs might consist of a sequence of daily closing prices of a stock, the temperatures at noon each day in your back yard, or the number of cases of measles treated in a hospital each day. The various noisy signals shown in Figure 6.1 are also examples of discrete data sets. (The data in that figure are the "dots." The lines connecting the dots are drawn to help us see the sequence in which the values occur.) When the input is discrete, the output from spectral analysis is also discrete.

Regardless of the nature of the input data, the output of spectral analysis—that is, the set of sine functions that have to be added together to make up the input signal—is usually summarized on a single graph. Along the x-axis of this graph is plotted frequency; along the y-axis is plotted a measure of the corresponding "energy." Displayed in this manner, the spectral analysis of a single sine function is a single spike located at the sine function's frequency with a height determined by the square of the sine function's amplitude. Figure 6.8 shows a typical example. The input for Figure 6.8 is a sine that repeats every 10 time units—that is, one that has a frequency of 0.1. Note that 0.1 is the only frequency in the figure for which there is a nonzero value along the y-axis. Similarly, the output of Figure 6.7 displayed as in Figure 6.8 would be three spikes—at the frequencies of the three sines shown and with heights determined by the respective amplitudes.

The y-axis in Figure 6.8 is labeled "intensity." Technically, intensity is a measure of the concentration of energy and the rate at which it is delivered. Intensity at a given frequency is roughly

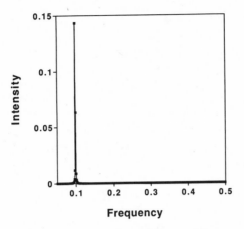

Figure 6.8 *The spectrum of a single, pure sine of frequency equal to 0.1.*

the "amount" of the input at that frequency. In the case of light, intensity is brightness; for sound, it is loudness. Because we are always interested only in relative intensities at different frequencies, the units of intensity are unimportant.

Note that the frequencies in the figure range between about 0.05 and 0.5. The frequencies are limited in this way because the input time series is a table of discrete data. Let's first see why 0.5 is the maximum frequency. Frequency is the reciprocal of the repeat time, so we ask, "What is the fastest the entries can repeat?" Well, they might repeat every unit of time, as the list 6, 6, 6, 6, . . . does, for instance. But such a set of data doesn't change at all; it's constant. A list of identical values is said to have zero frequency. (It has no wiggles, no variation). The next fastest repeat time is every two entries, as, for example, in the set 6, 3, 6, 3, 6, 3, ... The repeat time here is 2 units, so the corresponding frequency is 1/2, or 0.5. This is the highest frequency that can be associated with a list of discrete input data.

What about the low frequencies? Any real data form a set of finitely many entries. The lowest possible frequency we can assign to a finite list corresponds to a repeat time equal to the size of the list minus 1. For example, we can say nothing about the repeat time of the list 1, 2, 3. On the other hand, the list 1, 2, 3, 1 can be plausibly argued to repeat after 3 time steps. Of course, we can have only weak confidence in that claim. Our confidence would grow, however, if we saw the list 1, 2, 3, 1, 2, 3, 1, 2, 3, though the next entry *could* still be anything; it is not necessarily the expected 1. In general when we are performing spectral analysis, the bigger the data set, the better, and the more repetitions that occur in the set, the more confident we can be about the frequency of that repetition. Assigning the low-frequency components to a signal's spectrum is always chancy, so we suppress the lowest frequencies in all our figures.

The Spectra of Noise

Let's see what spectral analysis can tell us about noisy time series. Of the examples shown in Figure 6.1, Figure 6.1d seems to be the closest to repeating. Without being exactly periodic, it seems to harbor considerable undulation. The spectral analysis of the time series in Figure 6.1d appears in Figure 6.9. Indeed, there are only two sine components present — both with about the same amplitude. The ratio of the frequencies of these two sines is irrational — that is, the ratio cannot be expressed as a ratio of two integers. The sum of the two component sines never repeats in this case, but even so, such nonrepetitive behavior is neither random nor chaotic. It is rigidly deterministic; there is no chance anywhere. At the same time, it is not filled with unstable peri-

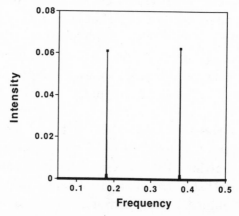

Figure 6.9   *The spectrum of the noisy time series shown in Figure 6.1d consists of two pure frequencies.*

odic behaviors and is not characterized by sensitive dependence on initial conditions. A nonrepeating signal whose spectrum consists of a few well-separated, incommensurate frequencies (frequencies with irrational ratios) is called *quasiperiodic*. Spectral analysis picks out quasiperiodicity immediately: the associated spectrum is a set of spikes.

The sum of two repeating lists with commensurate frequencies (frequencies whose ratio is rational) also repeats. As an example, add, term-by-term, the periodic lists

6, 3, 6, 3, 6, 3, 6, 3, 6, . . . , which has period 2

and

1, 2, 3, 1, 2, 3, 1, 2, 3, . . . , which has period 3

You get

7, 5, 9, 4, 8, 6, 7, 5, 9, . . . , a list with period 6

Contrast the spectrum shown in Figure 6.9 with those for time series with random elements, such as those in Figures 6.1a–c. Figure 6.10a shows the spectrum of the very erratic-looking time series in Figure 6.1a. Note that it contains no sharply defined spikes. We see a broad distribution of intensities at all frequencies. Similar broad distributions of intensities are observed in Figure 6.10b for the spectra of the somewhat less erratic-looking time series of Figures 6.1b (the open diamonds in Figure 6.10b) and 6.1c (the filled-in squares). Here, however, we see noticeably higher intensity at low frequencies and less intensity at high frequencies. The spectrum in Figure 6.10a is consistent with a horizontal line, a constant intensity at all frequencies. An erratic time series containing all frequencies in equal amounts is called *white noise* (in analogy with light, where a mixture of equal amounts of light of all colors is said to be white). A uniformly random time series, such as that shown in Figure 6.1a, is white noise.

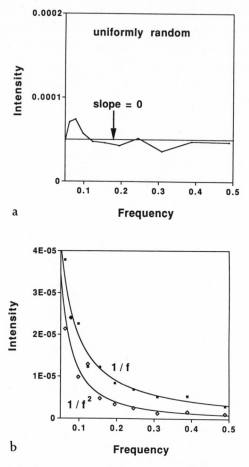

Figure 6.10   (a) The spectrum of the time series shown in Figure 6.1a —
white noise. (b) The spectrum of the time series shown in Figure 6.1b (open
diamonds) — Brownian noise — and the spectrum of the time series shown in
Figure 6.1c (filled squares) — 1/f noise.

The time series shown in Figure 6.1b is associated with a
random walk. We first encountered random walks back in Chapter Three when we discussed diffusion and diffusion-limited
aggregation. The random walk depicted in Figure 6.1b unfolds
along a line of length 1. The "state value" is the position of the
walker between 0 and 1 on this line. At any internal point, the
walker can lurch, with equal probability, a small distance every
time step toward the 0 end or the 1 end. At the end points 0 and
1, the walker encounters "barriers" and is "reflected" back into

the interior. Because only small steps are allowed, the walker's position at any instant is strongly correlated to its position in the previous instant. Incidentally, because such random-walking, diffusive motion was discovered by Robert Brown in the early nineteenth century, it is sometimes referred to as Brownian motion. The spectrum for Brownian motion reveals the strong correlation the signal has from instant to instant. If you ask, in Brownian motion, how much of a position change ("intensity") can be expected in a short time interval ("at high frequency"), the answer is surely "small." As the time interval is made larger (the frequency made smaller), however, the likelihood of large position changes (high intensity) increases. (Note that for the uniformly random time series, the likelihood of a large jump in state value is essentially the same for all time intervals.) It turns out that the Brownian spectrum is reasonably well fit by a smooth curve (see Figure 6.10b) of the form

$$\text{intensity} = \frac{\text{some constant}}{\text{frequency}^2}$$

For this reason, Brownian noise (noise with the same strong correlations as Brownian motion) is said to have a $1/f^2$ ("one over f squared") spectrum.

The time series shown in Figure 6.1c is like the noise observed in electronic devices to which we referred in Chapter One. Though visually not nearly so gently undulating as Brownian noise, this time series also does not seem so jaggedly irregular as white noise. Spectral analysis confirms this observation. Figure 6.10b shows that this time series also has a spectrum that falls off at higher and higher frequencies, but more slowly than Brownian noise. A reasonable fit to this spectrum can be obtained by a smooth curve of the form

$$\text{intensity} = \frac{\text{another constant}}{\text{frequency}}$$

For this reason, the time series shown in Figure 6.1c is said to have a $1/f$ ("one over f") spectrum. Noises with spectra that vary reciprocally with frequency, which are generically called $1/f$

noises—are everywhere. Examples, in addition to current fluctuations in electronic devices, include

Clumping of cars in traffic patterns on an interstate highway

Variations of electrical activity in the brain

Passage of sand grains through the neck of an hour glass or egg timer

Currents of sodium and potassium ions crossing the membrane of a living nerve cell

Variations in time keeping by quartz crystals

The approximately annual flooding of the river Nile

Short-term price changes in commodities and stocks traded in financial markets

Emphasis patterns in human speech

The large-scale distribution of matter in the universe (where, in this particular case, frequency means "how often in space" rather than "how often in time")

Though the question has been intensively investigated for many years, why $1/f$ noise should be present in so many different and seemingly unrelated systems is still unknown.

Let's summarize the preceding paragraphs: (1) Random time series have broad frequency spectra, not just isolated spikes. (2) Even if a time series is random, it may not be uniformly random; underlying rules may produce correlations between successive outcomes in a random time series. (3) A spectrum that is not constant implies that the associated time series has correlations. (That is, there are rules governing which randomly selected events can occur and which cannot—though, as in the case of $1/f$ noise, we may not know what the rules are).

The fully chaotic time series (one that takes on all values between 0 and 1) of the tent map (Figure 6.1e) and the logistic map (Figure 6.1f) are irregular but deterministic. They are exactly determined by dynamical rules. Can spectral analysis help us decide what these rules are? Figures 6.11a and 6.11b are the spectra of the fully chaotic tent and logistic maps, respectively! Neither shows sharp spikes at any frequency. Rather, we see broad distributions over essentially all frequencies. *Chaotic time series have broad (spread out, not discrete) frequency spectra* just like

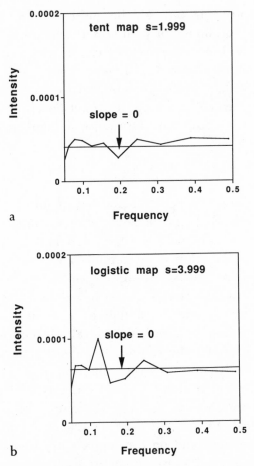

Figure 6.11   *(a) The spectrum of the tent map. (b) The spectrum of the logistic map.*

random time series. Note that in both cases shown in Figure 6.11, we have drawn horizontal lines as guides to the eye. The ripples in the distributions are statistical artifacts due to the finite size of the data sets used. Had larger sets been used, these ripples would have smoothed out. Both spectra are reasonably approximated by values that are independent of frequency; at each frequency there is about the same intensity. In terms of spectral analysis, at least, these chaotic time series are indistinguishable from white noise. There is no hint of the underlying rules from which these series emerged. Thus a nonconstant spectrum implies the existence of rules, but the existence of rules does not necessarily imply a nonconstant spectrum.

Spectral analysis is one of the most commonly employed techniques for probing noisy time series for hidden clues to underlying rules. Indeed, interesting information can be obtained by this technique. Generally, though, a good spectral analysis requires lots of data and some substantial computing capability. And even then, as we have seen, it cannot unambiguously differentiate between random causes for noise and deterministic chaos.

Delays

Producing pictures via spectral analysis requires a fair bit of technical wherewithal. In the next three sections, we describe different, and simpler, picture-making schemes for pursuing the same grail: the detection of determinism within noise. Each of the techniques we discuss attempts specifically to identify dynamical correlations between entries in a time series, in contrast to spectral analysis, which, though it is sensitive to *statistical* correlations, deals poorly with dynamical correlations.

Even when chaotic, the output of a deterministic dynamical system has rigid dynamical correlations; any entry in such a time series depends completely on the rules (the dynamic) and on the values immediately preceding it. In the output of a one-dimensional dynamical system, any entry (except the starting value) depends on the rules and on the single entry immediately preceding it. The tent and logistic maps are examples of one-dimensional systems. In either of those maps, we calculate the next value in the time series by applying the appropriate rule to the present value; we do not need to know any other values in the time series. Though all of science is predicated on the article of faith that adequate description of complex dynamical processes in nature often requires only a handful of variables, systems of a single dimension are rarely appropriate to the task. In a time series of one of the variables of an N-dimensional system, any entry depends on the rules *and* on the N immediately preceding entries. Consider as an example the two-dimensional Hénon map:

$$x_{n+1} = a - x_n^2 + b \cdot y_n$$
$$y_{n+1} = x_n$$

This map was derived by Michel Henon [96] to model a Poincaré map of Lorenz's weather equations. The Hénon map produces output that depends on two control parameters, a and b. More important, it generates two simultaneous time series — one in x, a second in y. Note that the second equation enables us to substitute $y_n = x_{n-1}$ into the first equation, yielding the equivalent rule

$$x_{n+1} = a - x_n^2 + b \cdot x_{n-1}$$

Thus for this map and any other two-dimensional map, the next x-value (x_{n+1}) depends on the current one (x_n) and the immediately previous one (x_{n-1}).

These remarks suggest a way of depicting the existence of dynamical correlations. Start with a time series arranged in a column. Place next to it an identical copy. Erase the first entry in the second column, and then shift the whole second column up one space. Label the first column x_n and the second x_{n+1}. Now treat each pair of entries in a row as the coordinates of a point on a graph plotted with x_n along the horizontal axis and x_{n+1} along the vertical — a first-delay plot. Here's a quick example; the time series is 1, 2, 3, 4, 5, 6, 7, 8, 9. In columnar format we have

x_n	x_{n+1}
1	2
2	3
3	4
4	5
5	6
6	7
7	8
8	9
9	—

Note that the nine entries we started with produce eight pairs. We then proceed to plot the points (1, 2), (2, 3), (3, 4), and so on.

If the time series originates from a one-dimensional dynamical system, the points so plotted all fall on some kind of curve. To get a clear picture of what the underlying dynamical rule is, we would like to fill in as much of the curve as possible. This is best accomplished if the time series is chaotic and is ranging over much of the possible extremes in state values. A periodic time series doesn't yield much of a graph. For example, if 1, 2, 1, 2, 1, 2, 1, 2, . . . is the time series, then (1, 2) is the first point, (2, 1) the second, (1, 2) the third, (2, 1) the fourth, and so on. In other words, a 2-cycle produces just two points on a plot of x_{n+1} versus x_n. You can fit anything from a straight line to the outline of a hippopotamus through two points. A better example is offered by this time series: 0.5, 0.99975, 0.00099950, 0.003993005, 0.01590426, 0.06258964, 0.23463004, 0.71813557, 0.80946506, 0.61677126, 0.94522152, 0.20705941. A plot of x_{n+1} versus x_n is shown in Figure 6.12a. The 12 entries are obtained from the logistic map with s = 3.999, and they range over much of the allowed region between 0 and 1. Because the points are well spread out, a smooth polynomial can be fit to the points, yielding the result

$$x_{n+1} = -3.999 \cdot x_n^2 + 3.999 \cdot x_n$$

That is, a delay plot of the data plus a smooth fit tells us exactly what the dynamical rule underlying the time series is. (Of course, we really need many more points than 12 to have a high confidence in the fit. See Figure 6.12b.)

What happens when the dynamics is of higher dimension than 1? That is, you have a series of values of a single variable, but the dynamics mixes that variable together with some unknown number of others. Let's take the Hénon map as an example. Figure 6.13 shows a first-delay plot of 1000 entries in a noisy time series of the x-component only (we don't know y exists, say) of the Hénon map. Figure 6.13 does not reveal a simple curve. Above some values of x_n lie more than one value of x_{n+1}; so x_{n+1} cannot be a simple function of x_n alone. The folded structure displayed in the figure implies that the dynamics is more complex than one-dimensional. (The folded shape of the Hénon

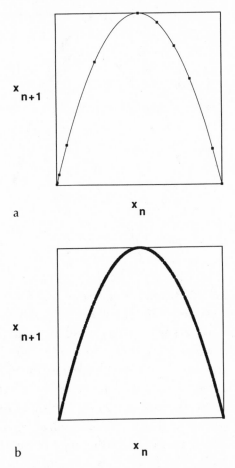

Figure 6.12   *First-delay plot of (a) 12 entries and (b) 1000 entries in a logistic map time series.*

delay map is actually a fractal. Successive blow-ups of a segment of the folds reveal more and more similarly shaped folds at higher and higher magnifications.)

Let's contrast Figures 6.12 and 6.13 with a delay plot of a random time series, as in Figure 6.14. One thousand entries from a uniformly random time series are used to make the (999) points plotted. Obviously, the uniform distribution of points over the plot is qualitatively different from a simple curve or even a folded cluster. The lack of structure apparent in the figure results from the fact that successive entries in a uniformly ran-

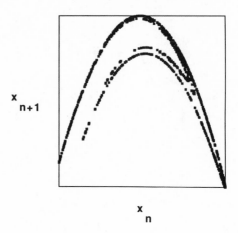

x_{n+1}

x_n

Figure 6.13 *First-delay plot of the x-component time series of the Hénon map.*

dom time series are totally uncorrelated. Any value can be followed by any other value; there are no rules. What do delay plots for random time series with correlations look like? Figure 6.15a shows a Brownian time series delay plot, and Figure 6.15b a 1/f time series delay plot. In Brownian motion, position at one instant is almost the same as position at the next instant. The entries in a time series with Brownian correlations change only gradually. A delay plot of Brownian noise consists of points whose horizontal and vertical coordinates are nearly identical — that is, of points closely clustered along the $x_{n+1} = x_n$ diagonal.

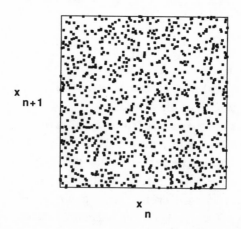

x_{n+1}

x_n

Figure 6.14 *A first-delay plot of a uniformly random time series.*

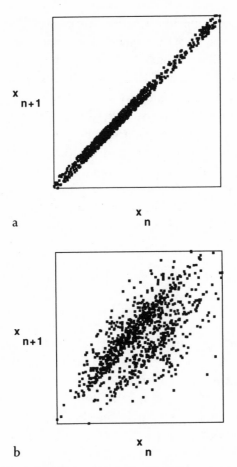

Figure 6.15   *(a) First-delay plot for a Brownian noise time series. (b) First-delay plot for a 1/f noise time series.*

The fuzziness of the clustering demonstrated in Figure 6.15a comes from the underlying randomness. The correlations in 1/f noise are intermediate between the strong correlations in Brownian noise and the absence of correlation in white noise. The delay plot for a 1/f time series shown in Figure 6.15b is a diagonal smudge, but it is much more widely dispersed than that for Brownian noise.

Let's summarize our observations up to this point. The first-delay plot for a periodic time series consists of a finite number of disconnected dots; the number of dots equals the periodicity of the series (two for period 2, three for period 3, and so on). If the first-delay plot suggests a simple curve, the time series is generated by a one-dimensional dynamical system (where x_{n+1} is a simple function of x_n). If the delay plot is a cluster with some sort of interesting geometrical structure, it is highly likely that the time series is generated by deterministic rules, but the dimension of the system is greater than 1. Random processes generate time series whose first-delay plots are smudges, though not necessarily uniform.

We have to be careful not to jump to conclusions about smudged delay plots. Randomness implies smudging, but smudging does not imply randomness. Here's an example. Imagine two logistic maps

$$x_{n+1} = s \cdot x_n \cdot (1 - x_n)$$
$$y_{n+1} = s \cdot y_n \cdot (1 - y_n)$$

where the starting values x_0 and y_0 are not the same. Even if x_0 and y_0 are close, when s is in the chaotic regime (near 4), the respective entries in the two time series will have little to do with each other after a while (the Butterfly Effect). Now suppose that what we measure is not the individual outputs, but the average, a_n, of the two:

$$a_n = \frac{1}{2} \cdot (x_n + y_n)$$

A first-delay plot of a_{n+1} versus a_n appears in Figure 6.16 for the case s = 3.999. Two features immediately appear: the points are smudged out, but the smudge has a sharply structured boundary. The geometrical pattern in the figure tells us that the time series we are dealing with comes from a dynamical system (structure in the blob) of dimension higher than 1 (not a simple curve). On the other hand, consider Figure 6.17. This figure shows a first-delay plot of the time series generated by averaging the output from six independent logistic maps, all starting at different initial values but all with s = 3.999. Note that the sharp boundary characterizing Figure 6.16 is now gone. Thus it is entirely possible to have a time series produced by deterministic rules (pro-

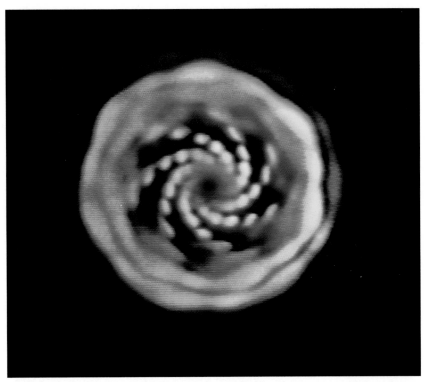

COLOR PLATE 1 Video color pinwheel.

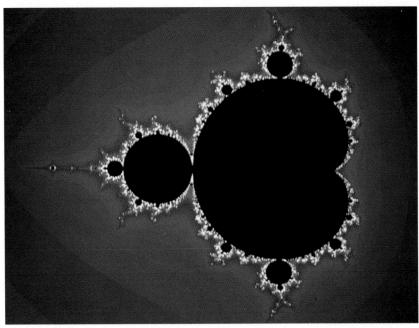

COLOR PLATE 2 The M-Set [Courtesy of A.G. Davis Philip].

COLOR PLATE 3   Color plate from the M-Set. Also see color plates 4 and 5. [Courtesy of A. G. Davis Philip]

COLOR PLATE 4

COLOR PLATE 5

COLOR PLATE 6 Color plate from the M-Set. Also see color plates 7 and 8. [Courtesy of Kerry Mitchell]

COLOR PLATE 7

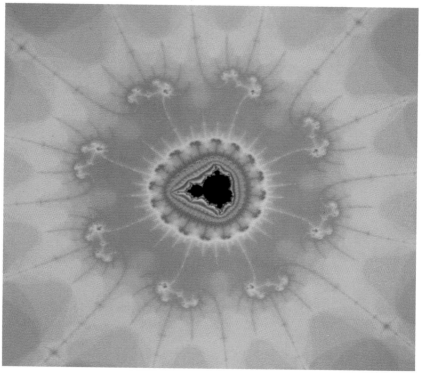

COLOR PLATE 8

COLOR PLATE 9 Rum Boat, 1983. [Courtesy of Elizabeth and Michael Rea]

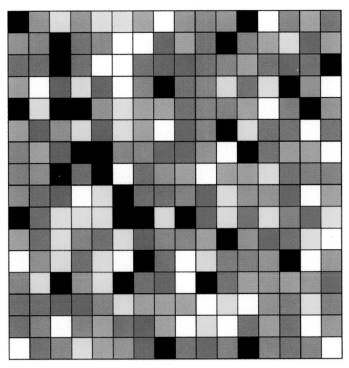

COLOR PLATE 10 Kelly white noise.

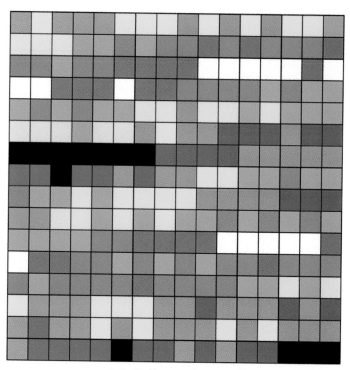

COLOR PLATE 11 Kelly random walk.

COLOR PLATE 12 Kelly 1/f noise.

COLOR PLATE 13 Cherry Tree.

COLOR PLATE 14 Zabriski Point. [Courtesy of Fenton K. Musgrave]

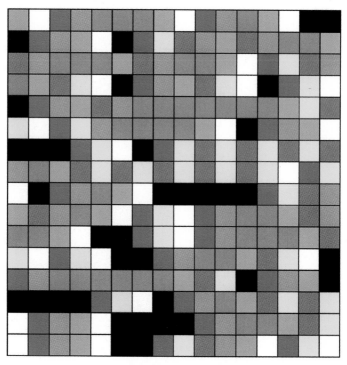

COLOR PLATE 15 Kelly tent map.

COLOR PLATE 16 Kelly logistic map.

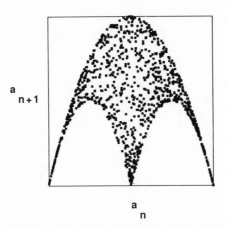

Figure 6.16 *First-delay plot of the average of two chaotic time series gener-ated by logistic maps starting at different initial values.*

vided there are enough of them) whose associated first-delay plot is a broad splotch showing no sharp geometrical substructure — just like that produced by a random process.

◆

Note that the points in Figure 6.17 are relatively tightly bunched, unlike the delay plot for uniform white noise. If we make a histogram of the underlying time series (that is, the one obtained from averaging the output of six independent logistic

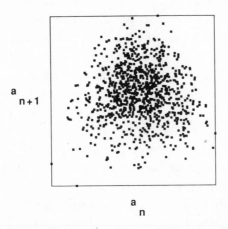

Figure 6.17 *First-delay plot of the average of six chaotic time series generated by logistic maps all starting at different initial values.*

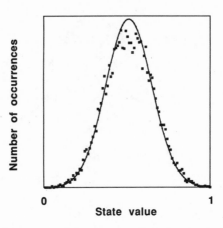

Figure 6.18   *A histogram of the average of six chaotic time series generated by logistic maps all starting at different initial values. The smooth curve is a fit to the data assuming a "normal" or "Gaussian" distribution.*

maps), as in Figure 6.18, we see the cause of this bunching. Only rarely will all six maps have outputs near 0 or 1, so only rarely will the average time series be near 0 or 1. Most of the time, the average output will be near 0.5.

The bunching seen in Figure 6.18 has an interesting shape. The smooth, bell-shaped curve in the figure is called a *normal* or *Gaussian* distribution. We find such distributions all around us. Plots of performances by a large number of people on skilled tasks (including SAT's, tests of reaction times, and times to run a mile) and plots of the physical characteristics (such as height, weight, or shoe size) of a large population yield examples of normal distributions. In fact, almost any repeated measurement on a physical, biological, or social system produces a range of outcomes. Most of these outcomes bunch around an average value, but some variations may be quite large—though the larger the departure is from the average, the less likely it is to occur. Often, the distribution of outcomes from repeated measurements follows a normal shape.

It is generally believed that a normal distribution is generated whenever the thing being measured depends in a simple way on a small number of more primitive parameters that can vary randomly (a person's weight might depend simply on bone size, muscle mass, and body fat, for example). A *Gaussian white noise* time series is a random sequence of outcomes whose histogram is shaped like a bell. Consider the following example: You have a large bucket filled with balls labeled 1 through 100. In the bucket is one ball labeled 1 and one labeled 100, one ball labeled 2 and one labeled 99, two balls labeled 3 and two labeled 98, . . . , sixteen balls labeled 16 and sixteen labeled 85, . . . , one hundred eight balls labeled 32 and one hundred eight labeled 69, . . . , and two hundred thirty-two balls labeled 50 and two hundred thirty-two labeled 51. Shake up the bucket and draw a ball; replace the ball, shake up the bucket, and draw another; repeat. Balls labeled 50 and 51 are most likely to be drawn; balls labeled 1, 2, 99, and 100 are least likely. After many, many trials, a plot of the number of times ball 1, ball 2, . . . , ball 100 have been drawn will be roughly a normal distribution. Note, however, that in any draw, any of the numbers between 1 and 100 *can* come up. The outcome of any draw is completely independent of the past (1 could come up ten times in a row, for example). Though we can make statements about trends or likelihoods for this process, we can never predict the next outcome; there are no correlating rules. The spectrum of Gaussian white noise is flat.

A delay plot of a Gaussian white noise time series is shown in Figure 6.19. We have chosen the shape of the histogram for this time series to coincide with the smooth normal distribution shown in Figure 6.18. Also, we have chosen values in the time series to be between 0 and 1 (instead of between 1 and 100 as in our previous example). Note that at first glance, this delay plot and the delay plot of the average of the six independent logistic maps (Figure 6.17) are indistinguishable. On closer inspection, however, we see a very subtle difference: though it is unlikely, 1 can follow either 0 or 1 in our Gaussian white noise series, whereas in the logistic map, an outcome near 1 is always followed by an outcome near 0 (never by another value near 1), and an outcome near 0 is always followed by another outcome near 0. (If that's not clear, make yourself a logistic parabola with s near 4. Then try starting with values near 0 and near 1 and see what you get.) Thus the upper righthand and lefthand corners of the Gaussian white noise delay plot will eventually get filled in

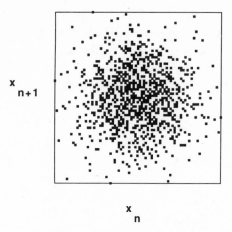

x_{n+1}

x_n

Figure 6.19   *First-delay plot of a Gaussian white noise time series. The shape of the histogram for this random series is the same as the smooth curve shown in Figure 6.18.*

(lightly) but will never get filled in by the average logistic output. You have to have a *large* data set to be confident about this difference.

As we have mentioned, the exact mechanism for producing that other commonly observed behavior, 1/f behavior, is not known, but now that we have the notion of a normal distribution, we can get an approximate feel for why so many 1/f distributions arise. The key issue is complexity and how a phenomenon is composed of its pieces. As we have said, measuring the heights of a large collection of people gives a "normal" distribution. Very roughly, this is because height depends on the length of leg bones and spine, and changing one of these by a small amount has a small effect on height. (Experiments on this topic were performed in Spain in the Dark Ages.) We say height depends *linearly* on the component phenomena, and a manifestation of the resulting normal distribution of heights is that although there may be people 50% shorter or taller than you, there certainly are not people 10 times taller or shorter than you (no one is 60 feet tall or 7 inches tall).

As a "counterexample," wealth depends in a very complex way on its component factors (wealth of parents, education, political connections, luck, . . .), and changing one of these by a small amount can have a large effect. This is because wealth depends *nonlinearly* on its components. Wealth has a lognormal distribution—that is, the logarithm of wealth is normally distributed. The distribution of wealth is obtained by exponentiation. Values near the mean get thrown far from the mean after exponentiating, and values far from the mean get thrown tremendously far from the mean. One consequence of this is that there are certainly people 10 times more wealthy than you, and (unhappily) people 1/10 as wealthy as you. Lognormal distributions are manifestations of the complexity of the phenomenon—of the degree of interrelation of its components. The more complex the phenomenon, the more the lognormal distribution comes to look like a 1/f distribution. Perhaps, therefore, inherent complexity is why so many things have 1/f distributions. (And perhaps not. Because no one yet knows how to quantify "complexity," this idea still awaits proof.)

How can we tell what the dimension of the process producing a time series is? Of course, if the first-delay plot yields a simple curve, the answer is that the process is one-dimensional. If the first-delay plot yields a smudge or complicated structure, however, the dimension of the process is higher than 1. One technique for uncovering the dimension employs longer delays. For example, from a time series $x_1, x_2, x_3, \ldots$, make all triples of the form (x_n, x_{n+1}, x_{n+2}). Plot these triples as points in a three-dimensional coordinate system. If the points lie on a simple surface, the dimension of the process is less than 3. If the points form a three-dimensional glob, then you have to try plotting quadruples $(x_n, x_{n+1}, x_{n+2}, x_{n+3})$ in a four-dimensional coordinate system. Of course, there's no way for us three-dimensional beasts to actually see four dimensions, so we have to rely on a numerical (and tedious) calculation of the dimension of the collection of points (*à la* Chapter Three) to visualize what happens. When a time series is produced by a deterministic process, the dimension so calculated stabilizes at some value as the dimension of the "embedding space" (the coordinate system in which the points are being plotted) is successively raised. The dimension of the embedding space at which this stabilization first occurs is said to be the dimension of the underlying dynamical system. The dimension of the delay plot of a randomly produced time series, on the other hand, is generally as large as that of the embedding space; a random time series usually produces a two-dimensional smudge in two dimensions, a three-dimensional smudge in three dimensions, a four-dimensional smudge in four dimensions, ... When the dimension is larger than about 4 or 5, identifying by delay methods the dimension of the dynamics causing a noisy time series requires a lot of data and a lot of computing power.

Close Pairs

As the chaotic output of a dynamical system evolves, it often occurs that two entries — perhaps well separated from each other in time — are quite close in value. Now in a one-dimensional dynamical system, the next output, x_{n+1}, is determined by the present output, x_n. Thus in a one-dimensional system, at least, the sequence of entries immediately following the first of the two close values will be similar to the sequence immediately following the second. (We'll get back to what happens in higher-dimensional systems presently.) Of course, because the two close values are not exactly the same, the sequences following them will diverge, eventually becoming nothing like each other (that's what the Butterfly Effect is all about). The closer the two "close" values are, the longer the two following sequences will stay close as well.

The appearance of approximately recurrent sequences in a chaotic time series suggests another graphical assay [97] for detecting underlying determinism:

Start with a time series.

Form all possible pairs of values.

Find the magnitude of the difference in the pair values for each case.

Select a "filter" value, and note which differences are less in magnitude than the filter.

A concrete example will help you work through the steps. Suppose the time series consists of four entries: 1, 2, 1, 3. With the first entry, 1, can be paired the second (2), the third (1), and the fourth (3). With the second entry can be paired the third and the fourth. (We've already noted the pairing of the second entry with the first entry.) With the third entry can be paired the fourth. This exhausts all the possible pairs. Now the magnitudes of the differences are

Pairs with the first entry: 1, 0, 2

Pairs with the second entry: 1, 1

Pairs with the third entry: 2.

Suppose we choose a filter of 0.5. Which pairs are closer than 0.5? Only the first entry paired with the third. On the other hand, if the filter value is 1.5, then the pairs with values closer than 1.5 are 1 with 2, 1 with 3, 2 with 3, and 2 with 4.

We would like to make a picture that readily shows this close-pair information and, in particular, gives evidence of strings of close pairs — because strings of close pairs are supposedly a signature for deterministic rules. A neat way to depict this information is to make a plot in which the horizontal axis is labeled "Entry number" and the vertical axis is "Delay." Here delay means the difference between the entry numbers of the pair members. Thus, entry number 1 paired with entry number 2 has a delay of $2 - 1 = 1$, and entry number 2 paired with entry number 4 has a delay of $4 - 2 = 2$. On such a plot, a dot is placed only at the coordinates of a *close* pair — a pair whose difference is less in magnitude than the filter value. A close-pairs plot for the short example given above with filter value equal to 1.5 appears in Figure 6.20.

A much more realistic close-pairs plot is shown in Figure 6.21, the plot for a 100-entry time series generated by the logistic map with $s = 3.999$; the filter is 0.01. First, note that the points plotted lie within a triangular pattern. That's because the deeper you go into the time series, the less the delay that's possible with the remaining pairs: 99 has only 100 left to be paired with, for example, because all other pairs involving 99 have been ac-

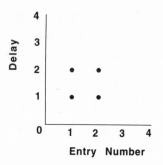

Figure 6.20   *Close-pairs plot for the time series 1, 2, 1, 3 with filter equal to* *1.5.*

counted for earlier in the analysis. Next, note that in addition to lots of isolated points, Figure 6.21 also seems to have many short horizontal strips. If we blow up the small boxed area in the lower left portion of the plot, as shown in Figure 6.22, we find that the horizontal strip contained within it consists of six points, all with the same delay value, 21. What this strip means is that in the given time series under analysis, entry number 2 is within 0.01 of entry number 23, entry number 3 is within 0.01 of entry number 24, . . . , and entry number 7 is within 0.01 of entry number 28. The entry string 2 through 7 is closely similar, member by member, to the entry string 23 through 28.

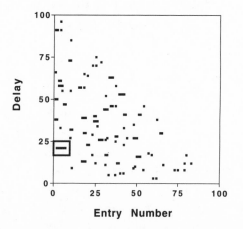

Figure 6.21   *Close-pairs plot for a logistic map time series with* $s = 3.999$ *and filter equal to 0.01.*

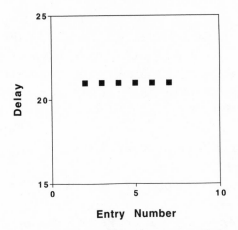

Figure 6.22   *Blow-up of the small boxed portion of the close-pairs plot of Figure 6.21.*

The existence of horizontal strips in a close-pairs plot is the signature for deterministic chaos. To see why, let's contrast the structure in Figure 6.21 with a close-pairs plot of 100 entries in a uniformly random time series with values between 0 and 1. Such a plot is shown in Figure 6.23. Nowhere in Figure 6.23 do we see horizontal strips of points. The points plotted in the figure scatter over the allowed triangular region and show no tendency to clump together. It is interesting to estimate how many close

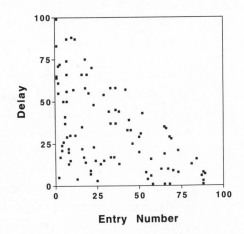

Figure 6.23   *Close-pairs plot for a uniformly random time series with filter equal to 0.01.*

pairs we can expect in a uniformly random time series. If the series consists of 100 entries, there will be $(100 \cdot 99)/2 = 4950$ pairs. (Imagine keeping tabs of pairs by labeling them with two entry numbers each. With each of the 100 first-entry numbers we can associate 99 different second-entry numbers. That makes a total of $100 \cdot 99 = 9900$ combinations. But the combination 3-7 is the same pair as the combination 7-3, so 9900 is twice the number of pairs.) If the filter for "closeness" is taken to be 0.01, then a second entry will be close to any given entry if its value is no greater than 0.01 more than that of the given entry and no less than 0.01 smaller. Around any given entry, then, there is a range of "closeness" 0.02 wide. Because the maximum variation any second entry can have relative to any first entry is 1, the probability of a close pair is 0.02/1 — that is, of every 100 pairs, about 2 will be close. There are 4950 possible pairs made of 100 entries, so we expect about $4950 \cdot 0.02 = 99$ close pairs. (Figure 6.23 actually has 95; that's chance for you.) Another way of saying this is to say that of the 4950 possible dots one could make on a close-pairs plot, only about 99 are actually made (only 99 pairs "pass through the filter"). What is the probability that two of the dots actually plotted will lie immediately adjacent to each other in a horizontal row? Because there are no correlations between the entries in the random time series, the probability of a horizontal strip of two dots is $(0.02)(0.02) = (0.02)^2 = 0.0004$. The probability that six dots will lie adjacent to each other in a horizontal row is the incredibly small number $(0.02)^6 = 6.4 \times 10^{-11}$. In other words, you can feel pretty secure that the row of six dots in Figure 6.22 didn't come about by chance.

Although close-pairs analysis works nicely for one-dimensional dynamics, it becomes less definitive as the dimension of the system increases. The reason for this degraded performance is apparent in Figure 6.24. You can see that in two dimensions, judging what "close" means requires information about *both* components of the dynamics. In the figure, x_1 is close to x_3 but y_1 is not close to y_3. The dynamics takes (x_1, y_1) into (x_2, y_2) and takes (x_3, y_3) into (x_4, y_4); x_1 and x_4 are very different. It's only when both x- and y-components of two points are simultaneously close that the following sequences will be close (as for points 3 and 6, 4 and 7, . . .). Because both components have to be close

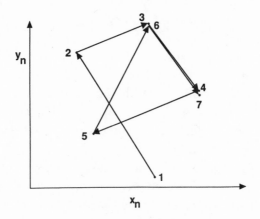

Figure 6.24   *A schematic two-dimensional dynamics.*

simultaneously, recurrences are less frequent in two-dimensional dynamics than in one-dimensional dynamics. We need more data to observe recurrent strings when the dynamics is two-dimensional. Also because both components have to be close simultaneously, strings diverge much quicker in two dimensions. The situation gets worse rapidly as the dimension of the dynamics increases. A close-pairs plot of a small time series when the dimension of the underlying dynamics is 6, say, almost never shows horizontal strips longer than about two. See Figure 6.25. It's pretty difficult to say whether Figure 6.25 is different from Figure 6.23.

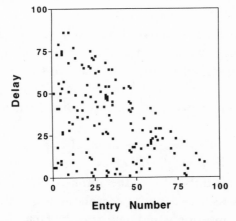

Figure 6.25   *Close-pairs plot of 100 entries in a time series produced by averaging the output of six independent logistic maps.*

Iterated Function Systems

Early in this book, we discussed image processing via iterated function systems and showed how one specific example — the Four Corner Chaos Game — could be used to uncover correlations between the bases in a strand of DNA. The Chaos Game can be used in a similar fashion to uncover correlations in a noisy time series. Here's how.

Start by breaking up the time series you're interested in into some number of bins. Though any number greater than two will generate an interesting picture, we prefer four, for reasons that will be clear in a moment. For simplicity, we scale all our time series such that the minimum value is 0 and the maximum value is 1. For every entry in the original time series, create an entry in a new series by the following simple rule:

If the original entry value is equal to or greater than 0 but less than 0.25, the entry in the new series is 1.

If the entry value in the original series is equal to or greater than 0.25 but less than 0.5, the entry in the new series is 2.

If the entry value in the original series is equal to or greater than 0.5 but less than 0.75, the entry in the new series is 3.

If the entry value in the original series is equal to or greater than 0.75, the entry in the new series is 4.

Use this new (coarsened) time series to select IFS rules.

Label the corners of a square 1 through 4, and take a starting point in the center of the square. Place a new point halfway from the corner determined by the first entry in the converted series to the starting point. Use the point so constructed as a new starting point, and repeat the last step with the next entry in the converted series.

You will recall that when the original time series is uniformly random, the collection of all points generated by the algorithm listed above uniformly fills in the interior of the square. When the original time series contains correlations, however, the filling of the square's interior is not uniform. In fact, that is precisely why we prefer four bins or four corners. Other numbers of corners will be filled by a uniformly random series in clumpy ways—sometimes with holes, sometimes with overlaps—that are characteristic of the number of corners. We don't want to confuse geometrical causes with correlational causes.

In any event, we present some of the possibilities. Figure 6.26 shows how other converted random time series fill the square. Note the corner labeling. A Gaussian white noise time series coarsened to entry values 1 through 4 contains more 2's and 3's and fewer 1's and 4's. A typical Gaussian string might be 3, 2, 3, 1, 2, 2, 1, 3, 2, 3, 4, 1, 3, 2, 2, 3, 3, and so on. Despite the fact that any value can follow any other, the IFS fill will not be uniform. More of the time will be spent near the 2 and 3 corners and near the edge of the square connecting those corners. See Figure 6.26a. Every feature in a simple IFS picture repeats on many different length scales, so the darkened vertical region along the 2-3 side of the square repeats again and again throughout the filled-in square.

Brownian noise that is converted to a sequence of the integers 1 through 4 has a very pronounced steplike character: frequently entries repeat, but when they change value, the change is almost always by one unit. A typical example is 2, 2, 2, 3, 2, 3, 3, 3, 4, 3, 3, 2, 3, 2, 2, 1, 1, 2, 1, 2, and so on. Because of this highly correlated stepping pattern, the IFS fill tends to wander between

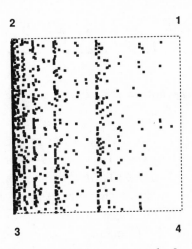

Figure 6.26a   *A four-cornered IFS picture made from coarsened Gaussian white noise.*

Figure 6.26b   *A four-cornered IFS picture made from coarsened Brownian noise.*

Figure 6.26c   *A four-cornered IFS picture made from coarsened 1/f noise.*

two adjacent corners for a while, then between two other adjacent corners, then between two others — but almost never between the corners 1 and 4. Thus the emerging fill is fairly well restricted to a thin crescent shape starting at 1, going to 2, then going to 3, and then ending at 4. See Figure 6.26b.

The correlations of 1/f noise are intermediate between uniform randomness and Brownian motion. Jumps between 1 and 3 (and vice versa) and between 2 and 4 (and vice versa) are more common than in Brownian noise, and jumps between 1 and 4 (and vice versa), though rare, are possible. Thus, in addition to the crescent backbone generated by numerous unit steps, there is some internal fill. The resulting IFS picture, Figure 6.26c, reveals these correlations.

So we see that the kinds of statistical correlations that spectral analysis picks up are also detected by IFS constructions. It's clear that the IFS technique senses something that spectral analysis doesn't: Gaussian white noise, like uniform white noise, has a flat spectrum but produces a nonuniformly filled IFS picture. What happens when we "drive" an IFS construction with deterministic noise? To see, consider Figure 6.27. Well, this is quite a surprise. Though the chaotic logistic time series appears spectrally white, it doesn't come close to uniformly filling in the IFS construction of a square. The very clear clumping of points harks back to our discussion in Chapter Two of the IFS picture made by a cyclic selection of the rules: a cycle forces a sequence of points plotted by IFS rules to land on top of each other again and again. A cyclic selection of the rules generates only small, separated islands of points. A cycle rigidly forbids certain occurrences. For example, in the 4-cycle 1, 2, 3, 4, 1, 2, 3, 4, . . . , a 2 always follows a 1; if a point has been plotted as a result of rule 1 having been chosen, then the next point can be plotted only in the "2 direction" from the last point. As a consequence, plotting points is strictly forbidden in much of the area defined by the square.

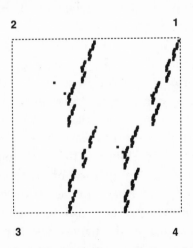

Figure 6.27   *A four-cornered IFS picture made from the chaotic output of the logistic map $x_{n+1} = 3.999x_n(1 - x_n)$.*

A similar prohibition results from chaotic dynamics. Here is a coarsened time series (it is read left to right and top to bottom) produced by a logistic map with s = 3.999:

```
2424124134113434341113424124134112434124241134241 1
2424134111241241242413434112424341134243424242412434
1241124113434124134243412434243413413412413413424243
4241113424242424124134243434343424241241124124 3434
2434124111241241342411343412424342412434134342434 1
2412424341342411134134124113411343413413413411 1 241
1242434242124113434341134111124243411341342424342 41
3411113413412413424341124111124113411112424242 4343
4111241243413413434241341111342424243411341242424 1
1243434241112424111124241113412434134424113413434 3
424111242434134112342434341243424124134 1
```

Careful inspection of this series reveals certain forbidden combinations. For example, 1 never follows 2 or 3, 2 never follows 2 or 3, 3 never follows 2 or 3, and 4 never follows 1 or 4. By dividing the square defined by the corners 1 through 4 into smaller subsquares and applying the arguments we developed in Chapter Two when we discussed cycles and the IFS generated by strings of DNA bases, you should be able to see how these forbidden combinations produce the clustering shown in Figure 6.27.

As the dimension of the underlying dynamics increases, the richness of alternatives also increases. (Recall Figure 6.24.) The IFS generated by a time series from a higher-dimensional system will fill more of the square. If, for example, we examine the coarsened time series generated by averaging the outputs of six independent logistic maps, we find that the only forbidden combinations from the preceding list that are still in effect are that 4 never follows 1 and 4 never follows 4. The IFS produced by such a time series is shown in Figure 6.28. Those portions of the IFS picture in this figure that are marked off by the small squares can contain points only when the time series used to generate the picture has a 4 following a 1 or a 4 following a 4 somewhere within it. (In the language of Chapter Two, these small squares have the strings 41 and 44 somewhere in their "addresses.") As

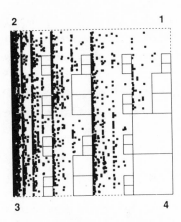

Figure 6.28   *A four-cornered IFS picture made from the average of the time series of six independent logistic maps (s = 3.999).*

you can see, all such regions in Figure 6.28 are empty despite the fact that the picture contains 5000 points. We previously pointed out that the time series created by averaging the outputs of six independent logistic maps shares many features (spectra, delay plots, close-pairs plots) with random, Gaussian white noise. Figures 6.28 and 6.26a (an IFS produced from Gaussian white noise) both have strong vertical stripes and appear at first glance to be very similar. Because a Gaussian white noise time series has no forbidden pairs, however, the IFS picture it generates can have points plotted in all of its subsections. We conclude, then, that given sufficient data, IFS can be a sensitive assay for deterministic causes in at least some noisy time series.

Before leaving this topic, we give an example taken from the real world. Spectral analysis and delay plots require a lot of data (many thousands of data points) to yield decent results, but IFS analysis often requires only a few hundred data points to give a quick sense of the nature of a time series. In Figure 4.4 we showed how the mean daily temperature varies over a year. The time series has a characteristic "backbone." Temperatures undulate seasonally in a sinelike fashion. More interesting is the nature of the short-term fluctuations. To investigate these more closely, we "detrend" the data—that is, we subtract the long-term behavior, leaving only the daily fluctuations. We fit a sine curve to the daily data and then, day by day, take the difference between the actual temperature and the value predicted by the fit.) We convert the resulting detrended noisy time series into strings of 1's, 2's, 3's, and 4's and plot an IFS picture from the strings. The results are shown in Figure 6.29. Despite the paucity of data, the plot is remarkably similar to Figure 6.26c—the IFS produced by 1/f noise. (What else?)

Noise, Music, and Visual Art

The technique of coarsening a time series into a string of integers to create IFS pictures suggests a couple of other ways to visualize correlations in noise by using colors and tones—both of which we introduced back in Chapter One.

Color selections for Plates 10–12, 15, and 16 were made according to the scheme shown in Figure 6.30. In each plate a 16 × 16 grid of small color squares was created using a different time series, where the first color entry was placed in the upper lefthand square, the second in the next square to the right, and so on. In other words, the time series can be read left to right starting from the top. (Recall that this way of rendering noisy time series was inspired by the work of the painter Ellsworth Kelly; see Chapter One.) Plate 10 shows what a uniform random

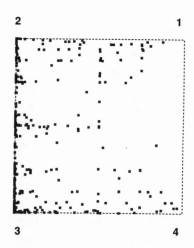

Figure 6.29   *IFS picture generated by daily temperature fluctuations.*

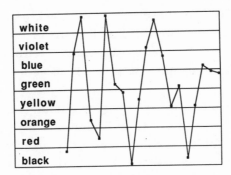

Figure 6.30 *A time series binned into discrete colors.*

time series generates: a random juxtaposition of the allowed colors—no rhyme or reason. The eye skips around, hunting without much success for patterns in this scrambled color landscape. Plate 11 is generated from a random walk. The sequentially banded structure of this plate clearly echoes the high degree of correlation in the time series used as its source. Plate 11 seems more soothing than the visual helter-skelter of Plate 10. Plate 12 originates from 1/f noise; though still evident, the banding here is much more broken than in the previous plate. The feeling of "energy" or "restlessness" stimulated by Plate 12 is intermediate between those produced by Plates 10 and 11.

Plates 15 and 16 have their roots in deterministic noise. Plate 15 is produced from the output of a single tent map, Plate 16 from the output of a single logistic map. At first glance these plates seem as disordered as if from white noise. They too are electric, busy, restless. (At the same time, in the palette choice of the plates, the predominant white component suggests a kind of Mondrian quality.) On closer inspection, we find short color themes whose insistent repetition almost certainly cannot be due to chance. (In Chapter Four we pointed out that chaos is filled with unstable periodicity. These recurrent, short color themes are a vivid affirmation of that claim.) Themes in Plate 15 include the frequent appearance of white-black-red, white-red-orange, and red-yellow-violet ripples, as well as long strings of blue following blue and long yellow-violet alternations. (Remember, the time series "wraps around": the last square in a row should be viewed as contiguous to the first square in the next row.)

Similarly, in Plate 16 there are white-black-yellow and green-white-black triples, as well as undulations of white and yellow, white and orange, and blue and violet. Within the complex juxtaposition of colors in these plates is an undeniable rigidity of structure.

In a similar manner, we can imagine binning a noisy time series into a set of discrete musical notes and *listening* for correlations. The perception of correlations can be enhanced by using the same noise source to assign different durations to the notes. Figure 6.31 shows a few different strings of notes. Even with varying durations, strings of notes originating from uniform white noise (Figure 6.31a) tend to have so little correlation — so little organization — that such sounds quickly become irritating. A Brownian motion time series is filled with correlation; its "music" (Figure 6.31b) wanders up and down like someone playing scales. These repetitive meanderings are too boring to hold our attention for any length of time. On the other hand, the intermediate correlations of 1/f noise actually generate note strings (Figure 6.31c) that harbor audio interest. The "music" of 1/f noise somehow balances familiarity and surprise, regularity and novelty. Finally, the "music" of the logistic map (Figure 6.31d) has much of the annoying, jagged irregularity of random noise (remember that the spectra of logistic-map chaos and white noise are indistinguishable), yet the careful listener will hear note

Figure 6.31   *Note strings generating from noisy time series. (a) Uniform white noise. (b) Brownian noise. (c) 1/f noise. (d) Chaotic output of the logistic map (s = 3.999).*

themes repeated again and again—just as color themes appear in Plates 15 and 16. Although color and note assays are formally identical, they seem different psychologically. That's probably because the whole record is always present in a painting, whereas music is ephemeral; you can scan a painting repeatedly to observe the correlations, but unless you have a trained musical memory, the information carried by tone sequences reverberates in your ears and brain for only a brief time and then is gone.

In Chapter One we mentioned that Richard Voss discovered a ubiquitous correlation in most real music (as well as in the emphasis patterns of human speech). His primary tool in this discovery was spectral analysis. When examples of real music are fed into a spectrum analyzer, often the resulting spectra vary with frequency like $1/f$. ("Stochastic" compositions, however, produce whitelike spectra. Of course, many would argue that random compositions aren't *real* music.)

Voss speculates that these observations bring an eloquent closure to one of the classical Greek theories of art. The Greeks believed that art imitates nature, and how this happens is relatively clear for painting, sculpture, and drama. Music, especially in the absence of accompanying words, was a puzzle, though. It had no obvious analogue in nature. Voss argues that the ubiquity of $1/f$ noise is the answer: music mimics the way the world changes with time. Somehow the composer, the performer, and the listener know this—even without a spectrum analyzer.

Alternatively, the connection between $1/f$ noise and music may be a kind of natural accident due to the statistical character of $1/f$ fluctuations. Martin Gardner [98] describes the situation this way?

It is commonplace in musical criticism to say that we enjoy good music because it offers a mixture of order and surprise. How could it be otherwise? Surprise would not be surprise if there were not sufficient order for us to anticipate what is likely to come next. If we guess too accurately, say in listening to a tune that is no more than walking up and down the keyboard in one-step intervals, there is no surprise at all. Good music, like a person's life or the pageant of history, is a wondrous mixture of explanation and unanticipated turns. There is nothing new about this insight, but what Voss has done is to suggest a mathematical measure of the mixture.

We can interpret Gardner's comment in the light of D. E. Berlyne's *Aesthetics and Sociobiology* [99]. Berlyne uses the notion of "arousal" as a way of understanding the nature of esthetic appreciation. In this case, arousal denotes the motivational level of emotions; emotions are aroused when they undergo a higher than usual level of activation. Of course, extremely high levels of arousal are unpleasant, so as the arousal level approaches the upper extremes, return to a lower level of arousal is pleasant. The amount of information transmitted by the stimulus is one of the factors that determines the arousal level. This amount of information depends in part on the subject. Some information may already be known, and some may fit into a more general matrix from which interrelationships (and hence redundancy) can be derived. Thus for an "uneducated" subject, the stimulus may contain so much information that it produces an uncomfortably high level of arousal and so is unpleasant. On the other hand, a stimulus that is thoroughly familiar to the subject may result in a low level of arousal because it carries very little information.

The information content is not the only contributing factor. For example, according to McClelland's theory of affect, any stimulus is compared with the subject's present adaptation level, which is a measure of the types of stimulation the subject has been receiving or is expecting to receive. Small departures in either direction from the subject's adaptation level produce a positive affect; larger departures produce a negative affect. In this way, affect reflects the level of novelty of a stimulus. Thus redundancy and novelty are both required for a pleasurable level of arousal.

Finally, there is speculation about a mechanistic, physiological reason why 1/f noise is more pleasing than the other kinds of noise. Gardner, in the same article from which we quoted previously, reports Mandelbrot as saying that noise at the point of sensory reception—the eyes, ears, and fingertips—is roughly white, whereas the closer it gets to the brain, the more closely it resembles 1/f noise. In other words, the nervous system seems to act as a filter to remove any part of the signal that is not 1/f. Voss speculates that the 1/f signal might be more pleasing than others because it more closely resembles the 1/f of our sensory experience.

Whether or not Kelly-type color checkerboards or musical note strings are useful for finding deterministic causes in time series of practical interest, the intricate contrasts and thematic repetitions created by different time series are visually and aurally appealing. Perhaps more interesting than using art and music to investigate noise is the potential for using noise to inspire art and music. Correlated noise may prove to be a rich source for constructing combinations of color, shape, pitch, and duration in which regularity and surprise are esthetically balanced.

Controlling Chaos

Okay, we've identified a deterministic component to a noisy signal. So what? Well, first of all, remember that chaos implies loss of only long-term prediction; short-term prediction is still possible. (At this writing, at least one company has been formed that purports to predict the short-term variation of stock prices on the basis of chaotic dynamics [100].)

But beyond prediction, any system that obeys rules — even if the behavior is chaotic — can be *controlled* once the rules are known. Recently, a potentially important harnessing of chaotic dynamics has been hinted at. As we've said before, within chaos are unstable periodic orbits: state values that, if they were supplied and iterated with infinite precision, would repeat over and over. Of course, any slight imprecision in these states causes successive iterates to wander away from the cyclic order and take on the complexity of the chaotic attractor. Nonetheless, in some simple situations the chaotic dynamics can be forced to track arbitrarily close to a specified unstable orbit by applying small, well-timed perturbations. An intriguing engineering implication of this idea is that it may be possible, by only the gentlest of teasing, to stabilize against undesirable chaotic vibrations structures (such as bridges, airplane wings, rocket engines, and the like) shaken by external influences. An at least equally important possibility is that of controlling a fibrillating heart by using the heart's own chaotic dynamics. Here's how "gentle" control of chaos works.

To start, one needs to know something about the underlying dynamics and the associated unstable cycles for the system of interest. If the dynamics is not known via a theoretical model, the necessary information can sometimes be obtained empirically from a chaotic time series by making appropriate delay plots. From such plots one can extract values of the state that are fixed or periodic without having to know the details of the rules generating the dynamics (in the best circumstances). Instead of examining a complicated actual system, let's see what the principles are by considering a familiar, fully worked out illustrative example, the tent map. (See Chapter Four.)

You will recall that the tent map has a single control parameter s that ranges from 0 to 2. When s is greater than 1, the tent gives chaotic output. For s between 1 and 2, the unstable, nonzero fixed point of the tent map has the value $s/(1 + s)$, so it lies between $\frac{1}{2}$ and $\frac{2}{3}$. To clamp the output of the tent to a string of approximately constant values, all near one of the allowed values of the unstable fixed point, we follow the procedure suggested in Figure 6.32. We start by choosing the appropriate s. That is, suppose we want a string of state values all close to 0.6. This value of the fixed point results when s = 1.5. We generate a time series for x = 1.5. Once the state of our system gets close to 0.6 (say to within 10%), we seek to keep it there despite its "desire" to wander chaotically away. Suppose that the state value

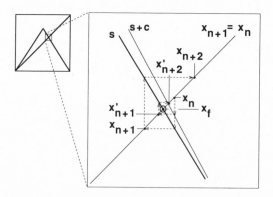

Figure 6.32   *Synchronizing the tent map to values near its unstable fixed point.*

is 0.63 — that is, less than 10% away from the desired value 0.6. The action sequence is initiated. For the sake of simplicity, we take action only when x_n is too large. (Other scenarios are possible.) The state value that follows 0.63 is $1.5 \cdot (1 - 0.63) = 0.555$, and the one after that is $1.5 \cdot (1 - 0.555) = 0.6675$. The latter value exceeds or tolerance level (0.66). If no action is taken, the succeeding state values are 0.49875 and 0.751875. That is, the distance of each successive iterate from the desired 0.6 grows rapidly. On the other hand, if at $x = 0.6675$, s is suddenly raised to 1.7 for a single iterate, then the succeeding state values become $1.7 \cdot (1 - 0.6675) = 0.56525$ and $1.5 \cdot (1 - 0.56525) = 0.652125$. Thus, by adjusting s once relatively slightly, we have managed to bring x below the critical value of 0.66. The strategy of perturbative control is to monitor x in each iterate and to increase x for a single iterate whenever x gets too big.

In the preceding example, the altered s was chosen to be 1.7 — out of the blue. How can we systematically go about choosing the change in the control parameter? We would like to find a way of changing it by small amounts and as infrequently as possible. The closer the corrective deflection is to the fixed point, x_f, the longer the system will go without needing additional corrections. A deflection such that x_{n+1} exactly equals x_f would necessitate no further correction, but such an infinitely precise deflection is impossible to achieve in practice. Here's a strategy for selecting a workable change in the control parameter.

Suppose we wait until x gets within plus or minus some *small* tolerance, t, of x_f. We do nothing until x_n is outside the tolerance limit. That is, we do not act until x_n equals $x_f + e$, where the magnitude of e (the "error") is greater than t. (We allow e to be either positive — x_n is greater than x_f — or negative — x_n is less than x_f.) If no action were taken, x_{n+1} would be given by

$$x_{n+1} = s \cdot (1 - x_n) = s \cdot (1 - x_f - e) = s \cdot (1 - x_f) - s \cdot e$$

or, because $x_f = s \cdot (1 - x_f)$, by

$$x_{n+1} = x_f - s \cdot e$$

Whereas x_n had an error of e relative to x_f, x_{n+1} has an error $-s \cdot e$, and, because s is greater than 1 (for chaotic behavior), this means that x_{n+1} is farther from x_f than was x_n.

On the other hand, suppose an additive correction (c) is applied to s between iterate n and iterate n + 1. In this case, the next value of x is

$$x'_{n+1} = (s + c) \cdot (1 - x_n) = s \cdot (1 - x_n) + c \cdot (1 - x_n)$$
$$= s \cdot (1 - x_f - e) + c \cdot (1 - x_f - e)$$
$$= x_f - s \cdot e + c \cdot (1 - x_f) - c \cdot e$$

It looks ugly, but writing x'_{n+1} like this suggests a slick trick. We choose the correction c such that $-s \cdot e + c \cdot (1 - x_f) = 0$ — that is, such that $c = s \cdot e / (1 - x_f)$. When this is true, $x'_{n+1} = x_f - c \cdot e$, and if the magnitude of c is less than 1, x'_{n+1} will actually be closer to x_f than was x_n. So can $c = s \cdot e / (1 - x_f)$ be less than 1 in magnitude? For the tent map, x_f is $s/(s + 1)$, so $1 - x_f$ is $1/(s + 1)$. Therefore, this choice for the correction c is equivalent to

$$c = s \cdot (s + 1) \cdot e$$

As long as the magnitude of the error is smaller than $1/[s \cdot (s + 1)]$, the magnitude of c will be less than 1. But how do we know that the magnitude of e will be less than $1/[s \cdot (s + 1)]$? The answer is that if x_{n-1} were no greater than t from x_f, then the magnitude of e would be no greater than $s \cdot t$. (That's because if $x_{n-1} = x_f - t$, say, then $x_n = x_f + s \cdot t$, according to the arguments given above.) If we require $s \cdot t$ to be less than $1/[s \cdot (s + 1)]$, then e will also be less than $1/[s \cdot (s + 1)]$. Thus the procedure outlined here will ensure that the time series stays within $s \cdot t$ of x_f as long as we apply the prescribed perturbation every time x gets further than t from x_f, provided that t is less than $1/[s^2 \cdot (s + 1)]$.

Synchronization to the unstable 2-cycle by applying a perturbation whenever the state is near x_2 can be accomplished by essentially the same argument. Thus, for example, if x_n is $x_2 + e$, then x_{n+1} is $x_1 - s \cdot e$. (Do you see why? Do the algebra.) A corrected s applied between iterate n and iterate n + 1 leads to $x'_{n+1} = x_1 - s \cdot e + c \cdot (1 - x_2) - c \cdot e$. (Again, do the algebra.) Choose c to cancel the term $-s \cdot e$. The result is $c = s \cdot e / (1 - x_2)$. But x_2 is given by $s^2/(s^2 + 1)$, so finally a correction for synchronization to the 2-cycle can be

$$c = s \cdot (s^2 + 1) \cdot e$$

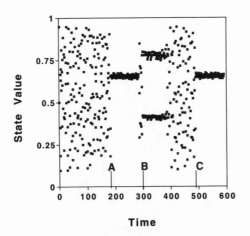

Figure 6.33 shows a time series of state values from the tent map. The uncorrected control parameter s is 1.9 for this case. The controlling sequence is initiated only when x gets to within a small tolerance of the fixed point, which in this case is 1.9/(1 + 1.9) = 0.655. . . . It takes almost 190 iterations from start-up for this to occur in the case shown. (Different start-up conditions would lead to different waiting periods—some short, some long.) Once x meets the criterion (at A in the figure), small perturbations are applied to the control parameter, only when needed, to keep x near the desired fixed point. After the control

Figure 6.33   *The chaotic output for a tent map with s = 1.9 is synchronized to the unstable fixed point at A, to the unstable 2-cycle at B, and back to the unstable fixed point at C.*

is turned off, the system runs freely again. At point B, x is sufficiently close to one of the two unstable 2-cycle values to begin synchronization to that behavior. Later, the system is switched back to fixed-point synchronization. The sequence of applied perturbations required for the synchronization to the fixed point is shown in Figure 6.34. You can see in Figure 6.34 that the control parameter need only be tweaked every few iterates to maintain the system near its unstable fixed point. The size of the perturbations shown are a maximum of about 5% of the size of the control parameter (that is, a maximum of about 0.1 out of 1.9). In short, it is possible in this simple mathematical experiment to flip back and forth between unstable periodic behaviors by using only small, infrequent perturbations of the system's control parameter.

Although the example discussed here depends on the familiar dynamics of the tent map, the ideas are more generally applicable. Indeed, the discussion in this section closely follows the report of W.L. Ditto, S.N. Rauseo, and M.L. Spano [101] on the first experiment on controlling deterministic chaos by small perturbations. These collaborators, motivated by the fundamental work of E. Ott, C. Grebogi, and J. Yorke [102], showed that it is possible to use the chaotic behavior of a relatively simple mechanical system to establish perturbative control without first

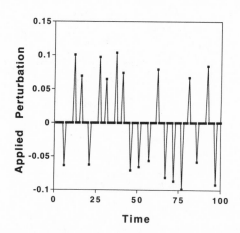

Figure 6.34   *The sequence of perturbations to the control parameter s = 1.9 that is required to synchronize the output of the tent map near its unstable fixed point.*

knowing the governing dynamical relationships. Of course, their experimental data produced a curved return map (a more probable occurrence than the nice straight lines of the tent). But synchronization to states sufficiently close to the desired state(s) leads to approximately linear dynamics (curves are approximately straight locally), and the guidance we extracted from the tent map remains remarkably appropriate for other, more realistic circumstances. Subsequent to the original experiment on mechanical oscillations, laboratory examples of controlling chaos in current oscillations in nonlinear circuits, in intensity variations in the light emitted by lasers, and in arrhythmia in beating rabbit heart cells have been reported [103].

Control of chaos in dynamical systems is in its infancy. We can expect it to be a focus of intense research for some time to come. It will be interesting to see what practical uses of these new notions emerge. Instead of engineering to avoid chaos, as has traditionally been the custom, it may be more productive to design systems to operate chaotically and exploit the order contained within chaos. Such systems have multiple, readily interchangeable, ordered (albeit unstable) behaviors. Such systems have the potential to adapt quickly to changing environments — to adapt on demand, so to speak. It will also be interesting to see if research finds that the trick of perturbative control has already been discovered by living organisms. Life often seems to operate on the brink of chaos, conserving structure on the one hand while encouraging variability and the exploration of diversity on the other.

Chaos and the Microworld

The microworld — the world of neutrons and protons and electrons, the world of the atoms from which we are all built — is not at all like the world we are accustomed to. Take the act of measuring an object's motion, for example. To measure the position of a car, say, you shine a little sunlight on it and look at the reflected light. The reflected light arrives at your eye, and you identify the car's position moment after moment, which provides you with position and motion information to great accuracy. Electrons are different. To determine where an electron is and how it is moving requires shining a little light on it, too, and detecting the reflections. Unfortunately, light packs a bit of a punch. The punch of sunlight on the motion of a car is totally ignorable, because the car is so large and the punch so feeble. But light bouncing off an electron knocks the electron about so badly that we lose our ability to say where it is and how it is moving. In the microworld, the effects of taking measurements are not benign and ignorable: measuring radically alters the thing being measured. This strong alteration of the state of the electron has an irreducible lower limit. There is every reason to believe that all measurements made on the microworld will produce fuzzy results. The best we can hope to do in specifying the motion of such objects as electrons is to construct statements of probability: If such-and-such measurement is repeated many times, the average value of all the outcomes will be so-and-so. The appropriate description of microworld matter—quantum mechanics—holds that processes at that level are infected with true randomness. Not just any old thing can occur in a measurement, but of the allowed possibilities, the one that actually appears (in any one measurement) cannot be predicted, even in principle.

Typical microworld energies are often at least 10^{20} times smaller than typical macroworld energies. Therefore, a large relative uncertainty for a micro-state is miniscule on the scale of a macro-state. As the energy of a microworld system is increased, the fuzziness of measurement and the infection of randomness become less important. Though it is ever present, the fuzziness of quantum mechanics eventually pales in comparison to macroscopic measurement uncertainties. The range of possible random outcomes, large in the microworld, is so compressed relative to macro-sized outcomes that we perceive no hint of quantum chance at our level. It is difficult to imagine how the macroscopic irregularities in the weather, the stock market, or our state of minds could be related in any way to the vagaries of the quantum realm.

But what about sensitive dependence on initial conditions for a microscopically chaotic system? Won't the uncertainties of the microworld be amplified into full-blown macroworld unpredictability for such systems? In principle, yes. But we have lots of more important uncertainties to worry about: that unaccounted-for flapping of a butterfly's wings is a vastly more significant source of error than any quantum fuzziness. In any case, on the macro-scale, chaos has an underlying attractor. Each little blip that tries to remove the chaotic system from the attractor's hold tends to get wiped out. Quantum randomness cannot be amplified into macroscopic randomness via chaotic processing. True randomness on the macro-level, if it exists, requires an alternative explanation.

And what, at the moment of this writing, is the status of chaos and/or randomness in everyday life? Regrettably—or, if you prefer high degrees of ambiguity, wonderfully—the answer is unclear. None of the tools developed so far for sniffing out determinism in noisy signals is powerful enough to distinguish high-dimensional chaos from true chance. Lots of macroscopic phenomena appear to have random features. But a string of numbers coughed out by a computer's pseudo-random number generator—a complicated, yet thoroughly deterministic, gadget—can seem pretty random too. As mentioned in the preceding section, we now know that some erratic physical devices can be synchronized into orderly behavior by tricks based on the properties of chaotic dynamics. Those examples offer compelling evidence that at least some subset of the noise of the macroscopic world is due to deterministic chaos. Indeed, with each new and physically different example of chaos under control, the odds that *all* macro-level noises have deterministic origins get a whole lot better.

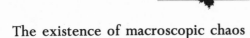

The existence of macroscopic chaos may have profound consequences for the structure of quantum mechanics. Though this is a very controversial point of view, here's the argument. On the macro-level a chaotic process cannot be "time-reversed" very far back. This means that if we measure a state to a finite degree of precision now, we will have more uncertainty about what state immediately preceded it, and even more about what state preceded that one, and so on. Eventually, the origins of the present state will be completely uncertain. This uncertainty as we try to construct all possible pasts of the present state grows much more rapidly ("exponential growth of uncertainty") for chaotic behavior than for regular behavior. Regular behavior can be time-reversed very much deeper into the past.

Quantum mechanics predicts that at the lowest energies, the states of all systems can be time-reversed. Pick a system that is known to be deterministically chaotic at macro-energies. At lowest energies it is time-reversible, whereas at highest energies it is not. How can time-reversibility get lost as the energy of the system is gradually increased? Joseph Ford, for one, maintains that it can't [49]. He argues that systems that are experimentally time-reversible in their lowest-energy states will also be experimentally time-reversible at highest energies. In consequence, he suggests, because the macro-behavior is known to be chaotic, the predictions of quantum mechanics cannot be complete. In his view, macroscopic chaos demands a new microscopic physics.

Despite its inherent, unappealing probabilistic character, quantum mechanics has withstood about three-quarters of a century of careful experimental examination. Its overthrow certainly will not be the result of theoretical arguments, however carefully constructed. The idea of such a revolution will be tolerated only when an experimental outcome — one not yet observed — demands it.

chapter

7

Benoit's World

INTRODUCED TO general readers on the cover of the August 1985 *Scientific American* [104], the Mandelbrot Set (see Plates 3–8 to refresh your memory) is now an element of popular culture. You can buy T-shirts and ties bearing it. You can send a postcard or tack up a poster of the Mandelbrot Set. You can find it on the jackets of CD's, on book covers, and in music videos. Magnifying tiny pieces of the Mandelbrot Set by computer is something of a cult fad. No one knows how many personal computers have been purchased worldwide for this (often unstated) purpose, but the number is surely not negligible. In many ways, the Mandelbrot Set is the prototypical fractal: so easy to generate, yet so complex in structure. Despite its algorithmic simplicity, the study of its regular irregularities has generated new and deep mathematics. In this chapter we connect the Mandelbrot Set to dynamics, describe how to draw it, and give a quick tour of a small fraction of its many fascinating and visually stunning features.

Before starting, we offer two cautions. First, unlike the topics of earlier chapters, little here is directly related to the natural world. The Mandelbrot Set will make a brief appearance again in Chapter Eight, but its main appeal outside mathematics is philosophical. It forces us to question the essence of our understanding of simplicity and complexity.

Second, this chapter is primarily descriptive. Much of the geometrical structure presented here is now understood analytically, but the mathematical tools involved are so advanced that we can't go into detail without greatly exceeding the scope of this book. However, we shall be careful to point out which results are established theorems.

A Transporter Machine

The Mandelbrot Set is intimately related to a two-dimensional dynamical system we call the Mandelbrot map, namely

$$x_{n+1} = x_n^2 - y_n^2 + a \qquad (1a)$$

$$y_{n+1} = 2 \cdot x_n \cdot y_n + b \qquad (1b)$$

(Though others studied some of the properties of equations (1a) and (1b) many years before him, the work of Benoit Mandelbrot over the past two decades has burned the importance of this dynamical system into our collective awareness.) You can think of equations (1a) and (1b) as defining a kind of science fiction transporter machine operating in two dimensions. At regular intervals in time, the space jumper — that is, the transportee — dematerializes at its current position, which is designated by coordinates (x_n, y_n), and rematerializes at a new position, designated by (x_{n+1}, y_{n+1}). The machine has one control knob to set values of a and another to set values of b. These knob settings serve to determine (though not directly) how far and in what direction the jump occurs. Each trip starts from the transporter machine located at $(x_0 = 0, y_0 = 0)$. Each knob can be set to either positive or negative values. (The sign helps ensure that all jump directions are covered.)

First, you can easily see that with the launch point at $x_0 = 0$, $y_0 = 0$, settings $a = 0$ and $b = 0$ lead to no jumps. Substituting 0's for x_0, y_0, a, and b in equations (1a) and (1b) produces $x_1 = y_1 = 0$ and so on. Consider the following nontrivial numerical example. The jumper launches from the origin $x_0 = 0$, $y_0 = 0$. The controls are set at $a = 1$, $b = 0$. Where does the jumper go under the given rules? Well, you compute x_1 by setting $x_1 = x_0^2 - y_0^2 + a = 0^2 - 0^2 + 1 = 1$. Similarly, $y_1 = 2 \cdot x_0 \cdot y_0 + b = 2 \cdot 0 \cdot 0 + 0 = 0$. Continue by setting $x_2 = x_1^2 - y_1^2 + a = 1^2 -$

$0^2 + 1 = 2$, and $y_2 = 2 \cdot x_1 \cdot y_1 + b = 2 \cdot 1 \cdot 0 + 0 = 0$. Repeat the process. Here's a short table of results:

n	x_n	y_n
0	0	0
1	1	0
2	2	0
3	5	0
4	26	0
5	677	0

(Check the entries to make sure there's no error.) Quickly, the jumper is zapped off along the positive x-direction — alas, never to return. The jumper is "lost in space."

Let's try another example to see what other possibilities there are. As before, $x_0 = 0$, $y_0 = 0$, only now $a = 0$ and $b = 1$. A short table of results in this case is

n	x_n	y_n
0	0	0
1	0	1
2	-1	1
3	0	-1
4	-1	1
5	0	-1

Now, instead of zapping off, the jumper quickly gets locked into a 2-cycle, one time landing on $(-1, 1)$, the next on $(0, -1)$, then back to $(-1, 1)$, and so on.

Okay, let's try one more with slightly messier numbers: $a = -1.5$, $b = 0$. In our first example, when b was also equal to 0, y started at 0 and stayed there. That's a general result. In each calculation of a new y, we sum two terms, the first term being a product with the old value of y as a factor, and the second term being b, which is 0 in this case. But if y starts at 0, then according to this rule it can never be anything but 0 (when b is 0 also). Thus we only have to calculate x's: $x_{n+1} = x_n^2 + a = x_n^2 + (-1.5)$ in this case. (The dynamics reduces to a one-dimensional system.) The results are

n	x_n	y_n
0	0	0
1	-1.5	0
2	0.75	0
3	-0.9375	0
4	$-0.6210 \ldots$	0
5	$-1.1132 \ldots$	0
6	$-0.2584 \ldots$	0

Indeed, after many more iterations (try some yourself), we find that the poor jumper rattles around in the interval between -1.5 and $+0.75$ on the x-axis without ever (exactly) revisiting any previous landing site. For these control settings, the sequence of jumps is bounded (the values remain finite) and chaotic.

We see in these examples, then, motion "escaping to infinity" (eventually), periodic cycling, and chaotic behavior—that is, a variety of interesting and quite different behaviors. It would be nice to have a user's manual for this transporter machine: a summary of all possible ultimate behaviors for all control knob settings. That is exactly what the bifurcation diagrams for the tent, logistic, and sine maps provided us: picture books of the dynamics' attractors, with one (one-dimensional) attractor picture for each (single) value of the control parameter. But displaying the bifurcation diagram for the Mandelbrot map will have to be done differently, because for the Mandelbrot map, each attractor is a set of points in the (x, y)-plane, and each control setting involves two values, an a-value and a b-value. To construct the bifurcation diagram of the Mandelbrot map exactly the same way as our previous diagrams would require attaching a two-dimensional attractor picture to each point in the plane made up of all (a, b)-pairs. We would need four dimensions to make such a diagram.

Because we have no way to create four-dimensional images, we settle for somewhat less information. Let's be content with knowing for each (a, b) no more than whether the resulting transport zaps off to infinity (lost forever) or stays bounded (a safe foray). The standard way of recording this more modest set of information is as follows: Imagine a picture of all possible control settings color-coded to indicate what settings are safe and what lead to escape. The control settings are laid out in a plane with coordinate axes: the a-axis in the horizontal direction and the b-axis in the vertical direction. Any point on this sheet identifies a pair of knob settings, (a, b). We refer to the sheet as the control space, or, interchangeably, as the (a, b)-plane. Now we select a point in the (a, b)-plane—that is, a pair of settings (a, b)—and run the machine. If the jumper does *not* escape to infinity in (x, y)-space (which we refer to as dynamics space) with that control setting, we color the point (a, b) black; if the jumper *does* escape to infinity, we color the point (a, b) some other color (we'll explain how to select the color in a moment). After we have tried every pair of settings, there will be a picture in the control space, some parts black, some parts other colors. All of the black points will be control knob settings for which the resulting transport does not escape to infinity. These black points are said to constitute the Mandelbrot Set. The Mandelbrot Set is the set of all "good" control values (that is, those that do not lead to escape) when initial launch occurs at the origin in dynamics space.

Of course, we face a computational dilemma: How long do we track the sequence of jumps before we decide on a coloring of the control set (a, b)? Waiting for the motion to escape to infinity would take an infinitely long time. Fortunately, the task is simplified by a useful theorem [105] that says that if the distance of the jumper from the origin ($x_0 = 0$, $y_0 = 0$) ever exceeds 2, the jumper will subsequently escape to infinity. For many values of (a, b), the jumper gets a distance of 2 from the starting point in a short time (see the first example above). Therefore, we adopt the following algorithm for coloring:

Draw a circle of radius 2 around the origin in the (x, y)-plane.
Pick a control set, a point in the (a, b)-plane.

Select a cut-off time — a maximum number of jumps — after which a coloring decision is made.

Run the transporter for the chosen control set. Keep track of the jumper's position [remember, that's in the [(x, y)-plane] and the number of jumps from the start.

If the jumper has not gotten to the circumference of the circle by the maximum number of jumps, color the point (a, b) black.

If the jumper gets to the circumference of the circle before the maximum number of jumps, color the point (a, b) a color chosen to indicate speed of escape — for example, violet for fast escape, red for slow escape, and colors in the order of the spectrum from violet to red for intermediate speeds. (The color coding is purely arbitrary; a different palette of colors may be more informative and also esthetically more interesting.)

Figure 7.1 shows a comparison of control space and dynamics space for the numerical examples discussed at the beginning of this section. Obviously, the point (1, 0) in control space does *not* get painted black, and the ultimate colors of the points (0, 1) and (−1.5, 0) remain in doubt after only three iterations.

In the coloring scheme described here, points near values of (a, b) that are colored black (no escape) are likely to be red (slow escape). Note, incidentally, that coloring a point black is a provisional statement about escape; it merely says that no escape has taken place within the cut-off time chosen. More of the (a, b)-plane is black if the cut-off time is low. For example, after a single jump (starting at $x_0 = 0$, $y_0 = 0$), $x_1 = a$ and $y_i = b$, regardless of what a and b are. Thus if the cut-off time was one jump, (x_1, y_1) would be inside a circle of radius 2 if and only if (a, b) was inside a circle of radius 2 centered on a = 0, b = 0. That is, the black portion of the control space is a circular disk of radius 2. Some of the control values associated with slow escape will always be incorrectly colored black, because a finite cut-off time will always be too short for some transport paths to get to the circumference of the radius-2 circle. There is less ambiguity as the cut-off time is increased. Figure 7.2 gives an idea of how the portion of the control space that is colored black shrinks as the cut-off time is increased. Approximate pictures of the Mandel-

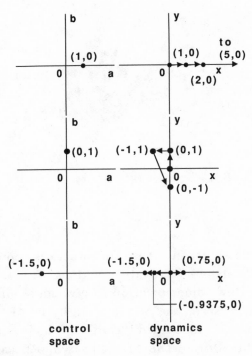

Figure 7.1   *Graphical representations of three iterations of the Mandelbrot map [in dynamics space, (x,y)] for each of three different control parameter values [in control space, (a,b)].*

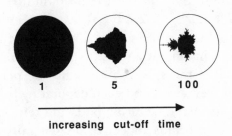

increasing cut-off time

Figure 7.2   *"Mandelbrot Sets" for 1, 5, and 100 maximum iterations. The circles in the second and third cases show the locations of the respective sets relative to the original disk of radius 2 centered on the origin.*

brot Set that are reasonably free of such errors require long cut-off times, so they take a long time to compute.

A Mandelbrot Set approximation with a fairly high cut-off time is shown in Figure 7.3. Superposed on the Mandelbrot Set

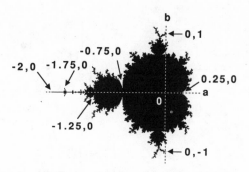

Figure 7.3   *The Mandelbrot Set, with some points labeled for reference.*

(the M Set for short) in Figure 7.3 are axes representing the
a- (horizontal) and b- (vertical) directions. The writing in the
figure notes a few values of (a, b) to give some information on
where in control space the M Set resides. The M Set is said to
consist of a main, heart-shaped ("cardioid") body decorated with
"buds" and "tendrils." The M Set is symmetrical above and
below the a-axis. The left-most point of the main cardioid is at
a = −0.75, b = 0, and the "heart's cleft" on the right is at
a = +0.25, b = 0. Attached to the main cardioid at a = −0.75,
b = 0, is a large bud, the "head"; the head, in turn, has a smaller
bud attached to it at a = −1.25, b = 0. After an infinite sequence
of ever-smaller buds, we see a long linear, but curiously decorated, tendril terminating at a = −2, b = 0. For obvious reasons,
this tendril is called the spike. All around the periphery of the
main cardioid, as well as around each decorating bud, are smaller
buds, each with its own decorating buds and tendrils. Though its
main body and largest buds appear to be generally roundish, the
boundary of the M Set is seen, under successive magnifications,
to be nowhere smooth. A careful examination of the attractors
of the dynamics of the Mandelbrot map associated with the
(a, b)-values displayed in Figure 7.3 shows that for every (a, b)
within the main cardioid, the sequence of transporter jumps in
dynamics space — which from now on we'll call *the orbit of* 0 —
converges to a fixed point. For every (a, b) within the head, the
orbit of 0 converges to a 2-cycle. For every (a, b) within the

largest bud on the head, the orbit of 0 converges to a 4-cycle. Once we leave these main buds and venture out into decorations whose points of attachment are off the a-axis, more interesting attractors arise. We will explore some of these in the next section.

A Quick Tour of the M Set

Figure 7.4 again displays the M Set, but a few interesting regions are enclosed in boxes. Subsequent figures examine the boxed regions under higher magnification.

Above and below the main cardioid—to its "north" and "south"—are reasonably large decorating buds. These have very noticeable tendrils or antennas protruding from them. Figure 7.5 magnifies one of these antennas (box *a* in figure 7.4). Perhaps most obvious in Figure 7.5 is that this antenna consists of *three* principal parts: a prickly trunk and two thorny branches making a sort of Y shape. On each of the branches are smudges that appear, at the level of magnification shown in Figure 7.5, to be similar to little bird's nests. A magnification of the "nest" on the left branch shows that it is really a small facsimile of the whole M Set! Such shrunken replicas are ubiquitous throughout the Set; they are called midgets. One of the fascinating aspects of the M Set is that although each midget resembles the shape of the

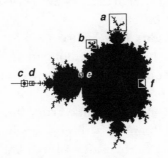

Figure 7.4   *A chart of the regions of the Mandelbrot Set to be magnified in the following figures.*

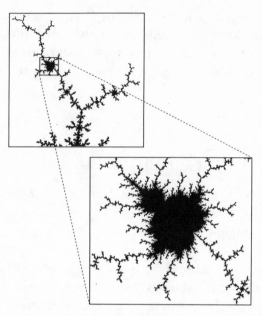

Figure 7.5   *A midget in the Y-antenna above the 3-cycle bud (box a in Figure 7.4).*

entire set, there are subtle differences. Another point to note is that each midget is itself surrounded by its own cloud of midgets, each of which, in turn, is surrounded by still smaller midgets, and so on forever. Thus the M Set is fractal-like, though of a much more intricate sort than we have seen before. The Gasket is not difficult to understand once you see its simple pattern; at each level you can predict exactly what you will see. The Mandelbrot Set, on the other hand, always reveals surprises, at least so far.

Back to the tour. Box *b* of Figure 7.4 encloses the antenna of the largest bud on the main cardioid to the west of the north bud. An enlargement of the box, as shown in Figure 7.6, reveals a structure with *five* primary segments. Now here is an amazing result: if one takes an (a, b)-value from inside the body of the north bud (whose antenna is a three-branched Y) and tracks the orbit of 0 for that (a, b), one ultimately obtains a closed *3-cycle* —that is, values of (x, y) in dynamics space that repeat after three complete iterations. Similarly, if an (a, b)-value from inside the body of the northwest bud (whose antenna is five-branched)

Figure 7.6   *The antenna above the 5-cycle bud (box* b *in Figure 7.4).*

is used, the orbit of 0 converges to a 5-cycle. The antenna structure attached to each bud attached to the main cardioid codes the period of the attractor to which the orbit of 0 converges for any (a, b)-value taken from the body of that bud. In fact, this pattern extends to the spike of the M Set: the spike can be considered a two-branched antenna (the two branches line up in the same direction) from the head, and for (a, b) in the body of the head, the orbit of 0 converges to a 2-cycle.

Figure 7.7 is a blow-up of box *c* from Figure 7.4, which surrounds the small blobby decoration far out on the main tendril. At the level of magnification of Figure 7.7, the blob reveals itself to be another midget. Values of (a, b) chosen from within the main body of this midget produce closed 3-cycle attractors for the orbits of 0. Similarly, magnification of the even smaller blob in box *d* of Figure 7.4 yields the tiny midget of

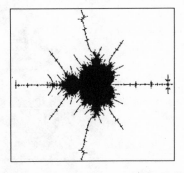

Figure 7.7   *The 3-cycle midget (box* c *in Figure 7.4).*

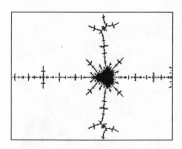

Figure 7.8   *A 5-cycle midget on the spike; there are two more (box d in Figure 7.4).*

Figure 7.8; (a, b)-values from its main body lead the corresponding orbits of 0 to 5-cycle attractors.

Figure 7.9, a blow-up of box *e* of Figure 7.4, shows something of how the main cardioid and the head of the M Set are joined. The slightly bumpy appearance (at the scale of Figure 7.4) of the

Figure 7.9   *Part of Seahorse Valley (box e in Figure 7.4).*

peripheries of these two large pieces of the M Set actually resolves into extraordinarily filigreed incursions into the region of control space where the Mandelbrot map produces unbounded behavior in dynamics space. Increased magnification of one of these filigrees shows a wonderland of intricately lacy swirls. Because of the suggestive shape of these filigreed decorations, this region of the M Set is often referred to as Seahorse Valley.

Box *f* of Figure 7.4 encloses the cleft of the heart. If we zoom in on the cleft, we see — in Figure 7.10 — what appears to be a parade of elephants emerging from its depths. This cleft is referred to as Elephant Valley. The elephants of this valley are not very robust creatures. In fact, they are incredibly wraithlike. At close inspection (see Figure 7.10 again), each is shown to be riddled with holes — a fragile, ghostly, doily fossil.

The final stop on this quick tour is at the very tip of one of the tendrils. (Except for the spike, it doesn't matter which one.) For specificity we examine the tendril attached to the southeast bud of the main cardioid, which, in frame *i* of Figure 7.11, appears to

Figure 7.10   *Part of Elephant Valley (box f in Figure 7.4).*

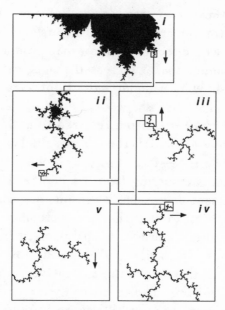

Figure 7.11 *A spiral antenna tip in the Mandelbrot Set. The winding behavior of the tip never ceases; the higher the resolution, the more the coil continues.*

be pointing due south. Closer scrutiny of this tip (frame *ii*) shows it to be pointing more westerly. But even closer inspection (frame *iii*) makes it appear to be pointing northward. You guessed it: in frame *iv*, at higher magnification, the tip points east, and then at still higher magnification (frame *v*) it's pointing south again. This spiral behavior occurs at the tips of tendrils everywhere on the periphery of the M Set.

The M Set is a wonderful teacher of humility. Although many general features of the M Set are understood now in a rigorous mathematical fashion, the surprises never end. In case by now you think every magnification of the M Set will yield midgets, seahorses, elephants, or branches, consider Figure 7.12, which shows two more magnifications of the M Set.

The M Set's surprises and intricacies are addictive. Indeed, the total computer time that professional mathematicians, computer scientists, and plain old hackers have spent probing and stretching the M Set is staggering. No other geometrical object has ever been subjected to such intense and universal scrutiny.

Figure 7.12   *Two more magnifications in the Mandelbrot Set.*

Some Geometry of the Mandelbrot Set

As intricate as all this structure seems, there is a pattern describing some of the organization of these features. First consider the decorations attached to the cardioid of the M Set. There are two *Principal Sequences* of buds — a northern and a southern sequence — around the cardioid, as illustrated in Figure 7.13. That is, the northern principal sequence (NPS) consists of the head (2-cycle bud), the large northern bud (3-cycle bud), the largest bud along the northern edge of the cardioid between the 3-cycle bud and the cleft of the heart (4-cycle bud), the largest bud between the 4-cycle bud and the cleft of the heart (5-cycle bud), and so on. The first few of these are labeled with their cycle numbers in Figure 7.13. The southern principal sequence (SPS) starts with the

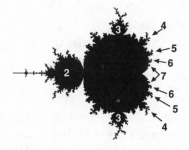

Figure 7.13   *Some bud cycle numbers in the North and South Principal Sequences.*

head and proceeds counterclockwise around the periphery of the cardioid.

Amazingly, a simple rule organizes all the rest of the buds attached to the cardioid. Recalling that the M Set is symmetrical above and below the a-axis, we consider only the NPS. Between the 2- and 3-cycle buds of the NPS, the largest is a 5-cycle bud. Between the 3- and 4-cycle buds of the NPS, the largest is a 7-cycle bud. Between the 4- and 5-cycle buds of the NPS, the largest is a 9-cycle bud. But more is true. Let's go back to the 5-cycle bud between the 2- and 3-cycle buds of the NPS. Between this 5-cycle bud and the 2-cycle NPS bud, the largest is a 7-cycle bud. Between this 5-cycle bud and the 3-cycle NPS bud, the largest is an 8-cycle bud. More yet: for example, between this 8-cycle bud and the 5-cycle bud, the largest is a 13-cycle bud. And so it goes. You see the general pattern: to find the periodicity of the largest bud between two given buds, just add the periodicities of the given buds. Between any consecutive buds in the NPS, the buds arrange themselves by size in a combinatorial pattern called a Farey sequence. The sequence is said to have different levels. In the following diagram, the first row is the first level, the second row the second level, and so on.

```
2 *   *   *   *   *   *   * 3 *   *   *   *   *   *   * 4   . . .
*   *   *   * 5 *   *   *   *   *   *   * 7 *   *   *   *     . . .
*   * 7 *   *   * 8 *   *   * 10 *   *   * 11 *   *         . . .
* 9 * 12 * 13 * 11 * 13  * 17 * 18 * 15 *                  . . .
```

With this pattern, we can identify the periodicity of any bud attached to the cardioid.

Moreover, this pattern of organization is repeated around each feature of the M Set, but relative to the cycle of the feature. For example, the head of the M Set is a 2-cycle bud. Attached to the head from west to east, the buds corresponding to the NPS have cycles 4, 6, 8, 10, 12, . . . —that is, $2 \cdot 2, 2 \cdot 3, 2 \cdot 4, 2 \cdot 5,$ $2 \cdot 6,$. . . . All the other buds (filled in according to the Farey sequence) that are attached to the head have cycle numbers twice those of the corresponding buds attached to the main cardioid.

The same is true for the buds attached to the 3-cycle bud of the NPS: the largest feature is a 6 (= 3 · 2)-cycle bud; those corresponding to the rest of the principal sequence have cycle numbers 3 · 3, 3 · 4, 3 · 5, 3 · 6, . . . ; and similarly for the others, filled in by the Farey sequence.

This same rule holds for the buds attached to all buds on the cardioid, for all buds attached to all of these buds, . . . , for all buds attached to every midget, for all buds attached to all buds attached to each midget. . . . There is an extension of this method that also gives the relative ordering of all the midgets [106]. In terms of the ordering of many of its pieces, the Mandelbrot Set is reasonably well understood.

We know more about the M Set. For example, despite the highly fractured structure of its boundary, the M Set is actually *one piece*. If the M Set were composed of separated pieces, then we could surround the pieces with shapes that would not touch and still enclose all of the Set within those shapes. *Every* attempt to separate the Set in this way fails, however. If you attempt to enclose all pieces of the M Set in nonoverlapping shapes, you always leave out some part of the Set [107].

Another surprising result is illustrated by several of the preceding figures. Simple two-dimensional objects, such as disks and squares, have edges that are simple one-dimensional curves. As we look more and more closely at the edge of the M Set, however, the picture seems to get fuzzier and fuzzier. In fact, the edge of the M Set is so fuzzy that its dimension is 2. This was proved by M. Shishikura [108]. The edge of the M Set is infinitely complicated.

Julia Sets

It's important to keep in mind that the Mandelbrot Set resides in control space. It is the collection of control settings for which the orbit of 0 in dynamics space does not escape to infinity. It's all of the safe control settings for transporter jumps starting at $x_0 = 0$, $y_0 = 0$. But dynamics space is closer to where the action is; that's where the jumper is zapping about. Suppose we place our transporter machine on a mobile unit and move the initial launch point around. Are all starting points equal? More precisely, suppose we set the controls for some well-known behavior when the starting point is at $x_0 = 0$, $y_0 = 0$. Will we observe the same behavior if x_0 and y_0 are not 0?

A moment's thought convinces us that the behavior does indeed depend on where the machine is set up. For example, suppose $a = 0$, $b = 0$, the "off" settings when $x_0 = 0$, $y_0 = 0$. But suppose $x_0 = 1$, $y_0 = 1$. What happens? Refer to equations (1a) and (1b). Plug in the starting values and proceed with the arithmetic. Here's a short table:

n	x_n	y_n
0	1	1
1	0	2
2	−4	0
3	16	0
4	256	0

Although $a = 0$, $b = 0$, produces no jumping when the start is at the origin, those settings lead to catastrophic jumping when the start is at $(1, 1)$.

If we are going to set up at different points in dynamics space, we are going to have to expand our user's manual to find out what the safe control values are. We need to know where we can set up our transporter for given control values without the jumper escaping to infinity. That is, what is the picture of all safe set-up places for each arrangement of control knobs? Because there are presumably many safe launch points for each control setting, and because there are many control settings, we'll need a pretty thick encyclopedia of pictures. Every such picture in our user's encyclopedia is called a Julia Set. More precisely, the set of all (x, y)'s in dynamics space whose orbits don't escape to infinity under the Mandelbrot map for a given control setting, (a, b), is said to be the *Julia Set* (technically, the filled-in Julia Set) for that (a, b). Curiously, Julia Sets are either connected globs or Cantor dusts in dynamics space. This was proved by P. Fatou and G. Julia in 1918–1919 [109]. Moreover, they knew exactly when each possibility occurred: if the orbit of 0 under equations (1a) and (1b) stays forever bounded, then the Julia Set for the corresponding (a, b) is connected; if the orbit of 0 escapes to infinity, however, then the Julia Set for that (a, b) is a Cantor dust. In the latter case, if the orbit of 0 escapes to infinity for some (a, b), there will still be set-up points in dynamics space for which the emerging orbit does not escape to infinity for that (a, b), but the safe launch points will be scattered and disconnected.

Despite Fatou's and Julia's theoretical understanding of the behavior of the dynamics defined by equations (1a) and (1b), pictures of these sets were known for only a few simple cases: the Julia Set of $a = b = 0$ is the unit disk; the Julia Set of $a = -2$, $b = 0$, is the interval $-2 \leq x \leq 2$ along the x-axis. The first fractal Julia Set was drawn *by hand* in 1925. Hubert Cremer [110] sketched a rough representation made up of triangles, and though only a few levels are shown, the basic shape of the Julia Set is apparent. To our trained eyes, the fractal nature of the Julia Set is apparent from this sketch. Nevertheless, it wasn't until 1978, when Mandelbrot applied the computer to the problem of constructing the M Set, that anyone had an inkling that relatively simple geometrical constructions could contain so many complicated and interesting shapes. Of course, as is true for just about everything associated with the Mandelbrot Map, its Julia Sets are fantastic. Figure 7.14 gives a very short and incomplete overview of some of the tamer Julia Sets of the Mandelbrot Map.

Though Figure 7.14 shows an outline of the M Set and some associated Julia Sets together in the same picture, remember that they "reside" in different spaces. Every point within the M Set is a different pair of control parameters (a, b) (for which the orbit of 0 is bounded), whereas every point within a Julia Set is a new (x_0, y_0), a new launch point for a bounded orbit. Clearly, if (a, b) comes from within the M Set (as in *a* through *e* in Figure 7.14), then $x_0 = 0$, $y_0 = 0$, is a good launch point, and it must lie within the corresponding Julia Set. The x-axis runs horizontally through the middle of each Julia Set, and the y-axis runs vertically through the middle of each.

Note that the Julia Sets that correspond to (a, b)'s chosen from the main cardioid are *single* blobs. For an (a, b) back near Elephant Valley (such as point *a*), the blob looks like a slightly puckered

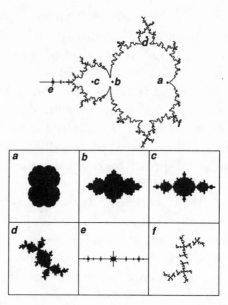

Figure 7.14   *Some Julia Sets and their corresponding points in or near the Mandelbrot Set.*

disk. In fact, at closer inspection, the puckers are very reminiscent of the heart cleft near where the (a, b) was chosen. As the M Set is traversed to the west (at *b*, for example), successive Julia Sets get ever more structured. The puckers become more pronounced and the Set appears to be pinching off into subsets. Once the head is entered (*c*), the pinches narrow down to single points—though the subsets remain connected to each other. Note that for (a, b)'s within the head, the orbits of 0 tend toward 2-cycles; not accidentally, the number of subsets joining at each pinch point is *two*. Up in the north bud (*d*), where 3-cycles attract the orbits of 0, the Julia Sets consist of pinched blobs where each pinch connects *three* subsets. The gross structures of the Julia Sets encode the character of the attractors for the corresponding (a, b)-values. As (a, b)'s close to the antenna are chosen, the corresponding Julia Sets become extraordinarily like the antenna itself. Out on the spike (*e*), Julia Sets take on a very spiky look. Finally, we observe that the Julia Set for an (a, b) outside of the M Set (*f*) disintegrates into a Cantor dust, though one with a structure strongly reflective of the boundary nearest that value of (a, b). In general, many (a, b)'s close to the boundary of the M Set have associated Julia Sets that look amazingly like that region of the boundary.

This last observation has been put on a rigorous footing by Tan Lei [111]. The boundary of the M Set contains certain (a, b)'s for which the orbit of 0 becomes exactly periodic. We ran through an example earlier: a = 0, b = 1. Recall that the orbit of 0 under the Mandelbrot map with this (a, b) is

n	x_n	y_n
0	0	0
1	0	1
2	-1	1
3	0	-1
4	-1	1
5	0	-1

Thus, in a finite number of steps, the orbit exactly repeats forever the values of $(-1, 1)$ and $(0, -1)$ and never returns to $(0, 0)$ or $(0, 1)$. In a finite time, the orbit becomes exactly periodic. (That's different from a situation in which an orbit asymptotically approaches cyclic behavior in an infinite number of steps.) We call such values of (a, b) preperiodic. (Another, not so informative name found in the literature is Misiurewicz points.) Suppose (a, b) is a preperiodic point. Magnify the M Set around this point. Now draw the Julia Set for this same value of (a, b). It is easy to show that (x = a, y = b) belongs to the Julia Set, so we can magnify the Julia Set around this point. When both are appropriately magnified around the point (a, b) (in two different spaces), the M Set and the Julia Set — for (a, b) — look more and more alike, possibly after a rotation. In Figure 7.15, this effect is illustrated for the preperiodic (a = 0, b = 1). This is not an isolated result: every disk drawn about any point of the boundary of the M Set contains infinitely many preperiodic points. For all these points, the M Set gives sort of a picture of the Julia Sets of those points. This has led some to say (in a slight oversimplification) that the M Set is itself a one-page encyclopedia of the Julia Sets.

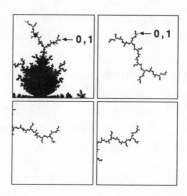

Figure 7.15   *An illustration of Tan Lei's theorem. Top left: a portion of the Mandelbrot Set with the point (a = 0, b = 1) indicated. Top right: the Julia Set for (a = 0, b = 1) with the point (x = 0, y = 1) indicated. Bottom left: a magnification of the Mandelbrot Set about (0, 1). Bottom right: a magnification of the Julia Set about (0, 1).*

Recoding the Logistic Map

You may well ask, "What is so special about equations (1a) and (1b)?" Why not choose other, perhaps simpler, transport rules? Though not immediately obvious, the answer is that equations (1a) and (1b) are direct consequences of generalizing the logistic map. In some sense, the Mandelbrot map *is* the logistic map. Let's see how this claim can be true.

If we set b = 0 in equation (1b) and take $y_0 = 0$, then equation (1a) reduces to the one dimensional map

$$x_{n+1} = x_n^2 + a \qquad (2)$$

This is equivalent to a transport machine in one dimension, beaming the jumper along the x-axis only, with a single control knob, a. Recall that the logistic map is

$$x'_{n+1} = s \cdot x'_n \cdot (1 - x'_n) \qquad (3)$$

The reason why we write x' in the latter equation is to remind us that at this point, at least, there is no connection between iterates of the Mandelbrot map and those of the logistic map. The iterates of the logistic map are supposed to represent fractions of a maximum population size; they are supposed to range between 0 and 1. Even so, we can switch to the transporter machine metaphor for the logistic map as well. In this language, the logistic map can be viewed as a one-dimensional transporter, beaming the jumper along the x'-axis, with a single control knob, s. Remember that when s is between 0 and 4, the iterations of the logistic map that start between 0 and 1 remain bounded between 0 and 1. When s is greater than 4, however, almost every starting value careens off to $-\infty$, a result that is impossible to reconcile with populations (what population is negative?) but easy to imagine in terms of transports. The exceptions form Cantor Sets (recall, for example, "Dust in the Tent" in Chapter Four). Figure 7.16 shows a case where $s > 4$ and where typical starting values head in short order toward the nether world. (Note that we have to extend both the 45° line and the logistic parabola into the regions to the left of $x = 0$ and to the right of $x = 1$ in order to complete the graphical iteration.) Thus, like the Mandelbrot machine, the logistic machine has both "good" and "bad" control settings. (And exactly as in the Mandelbrot case, when the settings are "bad," there are dusts of points along the x'-axis from which launches are safe.)

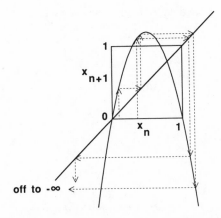

Figure 7.16 *Escaping to $-\infty$ for the logistic map with $s > 4$.*

To show that the two maps in the preceding paragraph contain identical information, we make a translation from logistic language into Mandelbrot language. That is, we seek a coordinate transformation taking us from x' to x. Because both maps are quadratic (the highest power on the righthand sides is 2), we try the simple transformation

$$x_n = A \cdot x'_n + B \tag{4}$$

Our task is to find the values of A and B that ensure that equations (2) and (3) remain simultaneously valid. First we plug equation (4) into equation (2) everywhere we see an x:

$$A \cdot x'_{n+1} + B = (A \cdot x'_n + B)^2 + a$$

Then we rewrite this with x'_{n+1} alone on the lefthand side:

$$x'_{n+1} = A \cdot x'^2_n + 2 \cdot B \cdot x'_n + \frac{B^2 - B + a}{A}$$

and compare the result with equation (3) written out as

$$x'_{n+1} = -s \cdot x'^2_n + 2 \cdot x'_n$$

These last two equations are supposed to be the same, and the only way they can be, for all values of x'_n, is if the stuff multiplying x'^2_n in the two expressions is the same:

$$A = -s$$

the stuff multiplying x'_n in the two expressions is the same:

$$2 \cdot B = s$$

and

$$\frac{(B^2 - B + a)}{A} = 0$$

In short, equations (2) and (3) are identical if

$$x_n = -s \cdot x'_n + \frac{s}{2}$$

and

$$a = \frac{s}{2} - \frac{s^2}{4} \qquad (5)$$

Let's make a short table of translation from s-values to a-values:

s	a
0.00	0.00
1.00	0.25
2.00	0.00
3.00	−0.75
4.00	−2.00

Figure 7.17 shows plots of equation (2) for several different values of s. Equation (2) defines a family of parabolas, the mem-

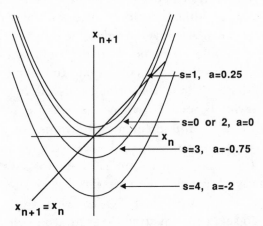

Figure 7.17   *Graphs of equation (2) for several s- (and a-) values.*

bers of which are identical except for where their minima are positioned along the $x_{n \cdot +1}$-axis (that is, the vertical axis). For any parabola in the family, the minimum value of x_{n+1} occurs where x_n is 0 (otherwise x_n^2 is positive). That minimum value, therefore, is just a. The dynamics of the logistic map is controlled by the single parameter s; the dynamical system defined by equation (2), the x-part of the Mandelbrot map, has the single control parameter a.

It's instructive to draw some graphical iterations for the map in equation (2). Of all the starting values between 0 and 1 in the logistic map, $x'_0 = \frac{1}{2}$ escapes to infinity the fastest. (That's the value of x' sitting directly under the top of the logistic parabola.) To what value of x in equation (2) does this choice correspond? The answer is

$$x_0 = -s \cdot \left(\frac{1}{2}\right) + \frac{s}{2} = 0$$

Thus the most economical exploration of escape in the logistic map is the orbit of $\frac{1}{2}$, and in the Mandelbrot map, the orbit of 0. (Besides economy, these orbits are chosen for subtle technical reasons related to the theorem of Julia and Fatou that determines which Julia Sets are connected and which are Cantor dusts.) Figure 7.18 depicts orbits of 0 for several possible values of a. For a value of a that is intermediate between $+\frac{1}{4}$ and -2, all of the various cyclic and bound chaotic behaviors discussed previously for the logistic map are evidenced. One example, in (i) (for a = 0), shows a stable fixed point attracting the orbit of 0; a second, in (ii), shows the orbit converging to a stable 2-cycle. In (iii), a is less than -2, and the orbit heads off to ∞. The same divergent behavior is seen in (iv), where a is greater than $+\frac{1}{4}$.

The short table we found for relating values of a and s indicated that a ranges between -2 and $+\frac{1}{4}$ when s ranges between 0 and 4. Cases (iii) and (iv) of Figure 7.18 are for a-values that lie outside the range for which we have familiar s-values. Suppose a is less than -2 or greater than $+\frac{1}{4}$. To what s-values do these correspond? To see, turn equation (5) around and write s in terms of a:

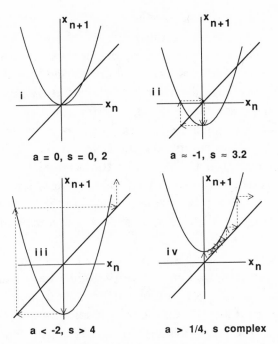

Figure 7.18 *Graphical iteration of equation (2) starting from $x_0 = 0$, for several values of a.*

$$s = 1 \pm \sqrt{1 - 4 \cdot a} \tag{6}$$

Equation (6) tells us that for every value of a, there are two possible values of s that could have given rise to that a (except for $a = \frac{1}{4}$, where $s = 1$ only). When a is negative, s has two real values, one positive, one negative. (A *real number* is a number whose square is positive.) Again, a negative s-value for the logistic map as a model of population dynamics poses interpretative problems. But such a value is fine if we think of equation (3) as a transporter machine. When a is between 0 and $\frac{1}{4}$, s has two positive real values. If $a > \frac{1}{4}$, then $4 \cdot a > 1$, and the value inside the square root of equation (4) becomes negative; $\sqrt{1 - 4 \cdot a}$ is consequently a number whose square is negative. Any such number is said to be *imaginary*. A number that is the sum of a real number and an imaginary number — such as in equation (6) when

$a > \frac{1}{4}$—is called a *complex number*. Thus panel (iv) of Figure 7.18 corresponds to a complex value for s!

Real a-values greater than $+\frac{1}{4}$ are associated with complex values of s. Only very special complex values of s lead, through equation (5), to real values for a. In the spirit of exploration, we might inquire what the orbit of 0 under equation (2) looks like for the other, nonspecial complex values of s. General complex values of s produce complex values for a (from equation 5) and for x_n (from equation 4). To answer the original question that we set out to investigate—namely, where the Mandelbrot map, equations (1a) and (1b), came from—and to round out this discussion, we need to know a few properties of complex numbers.

Here's a brief list of things we need to know. As we said before, a complex number can be thought of as a sum of a real number and an imaginary number; it's a kind of binomial. The imaginary part of a complex number is sometimes expressed as a product of a real number times the unit imaginary number i. The latter quantity has the property that its square is -1:

$$i^2 = -1$$

Thus if a and b are real, then $c = a + b \cdot i$ is a complex number. Now suppose $d = e + f \cdot i$ is a second complex number (that is, e and f are real). Here are some essential properties involving c and d:

Equality. Two complex numbers are equal if and only if their real parts are equal and their imaginary parts are also equal. That is, c and d are equal if and only if $a = e$ *and* $b = f$.

Addition. The sum of two complex numbers is also complex, with the real part equal to the sum of the constituent real parts and the imaginary part equal to the sum of the constituent imaginary parts. Thus the sum of c and d is defined by

$$c + d = (a + e) + (b + f) \cdot i$$

Here's an example. Let $c = 2 + 3 \cdot i$ and let $d = 5 - 2 \cdot i$. What is $c + d$? The answer is

$$c + d = (2 + 5) + [3 + (-2)] \cdot i = 7 + (1) \cdot i = 7 + i$$

Multiplication. The product of c and d is obtained by multiplying the two as though they were binomials (via the FOIL method, for example).

$$c \cdot d = (a + b \cdot i) \cdot (e + f \cdot i)$$
$$= a \cdot e + a \cdot f \cdot i + b \cdot i \cdot e + b \cdot i \cdot f \cdot i$$
$$= a \cdot e + (a \cdot f + b \cdot e) \cdot i + (b \cdot f) \cdot i^2$$

But because $i^2 = -1$, we find that

$$c \cdot d = (a \cdot e - b \cdot f) + (a \cdot f + b \cdot e) \cdot i$$

Here's another example: Take c and d from the previous example. Then

$$c \cdot d = [(2)(5) - (3)(-2)] + [(2)(-2) + (3)(5)] \cdot i$$
$$= 16 + 11 \cdot i$$

Squaring. An immediate consequence of the product rule is that the square of a complex number, c, is

$$c^2 = (a + b \cdot i)^2 = a^2 - b^2 + (2ab) \cdot i$$

Thus, using c from the previous examples, we get

$$c^2 = (2^2 - 3^2) + (2 \cdot 2 \cdot 3) \cdot i = -5 + 12 \cdot i$$

Though there are many additional properties of complex numbers, the only one we shall need here is the magnitude.

Magnitude. The magnitude of $c = a + b \cdot i$ is the positive number $\sqrt{a^2 + b^2}$.

Again, using c from before, we find that

$$\sqrt{2^2 + 3^2} = \sqrt{13}$$

and the magnitude of d is

$$\sqrt{5^2 + (-2)^2} = \sqrt{29}$$

Finally, and very important for what we want to do, a complex number can also be represented as a point in the complex plane. The *complex plane* is a Cartesian, x-y coordinate system in which real values only are plotted along the x-axis and imaginary values only are plotted along the y-axis. Thus, in this representation, the complex number $z = x + y \cdot i$ is the coordinate pair (x, y). Figure 7.19 shows four examples of complex numbers: $A = (2, 7) = 2 + 7 \cdot i$, $B = (-6, 13) = -6 + 13 \cdot i$, $C = (-6 - 4 \cdot i$, and $D = (4, -3) = 4 - 3 \cdot i$. Actually, the complex plane is just the real plane plus a prescription for multiplying two points to obtain a third point. That is, the product of the two points A and B from Figure 7.19 is the new point $(2 + 7 \cdot i) \cdot (-6 + 13 \cdot i) = [2 \cdot (-6) - 7 \cdot 13] + [2 \cdot 13 + 7 \cdot (-6)] \cdot i = -103 - 16 \cdot i = (-103, -16)$.

Using this representation of complex numbers as points in the plane, and recalling Pythagoras's Theorem, we see that the magnitude of a complex number z is just the distance from the corresponding point to the origin. Because the absolute value of a real number can be viewed as the distance from the number to 0, the magnitude of a complex number is a generalization of the notion of absolute value. Consequently, the magnitude of a complex number z is often denoted $|z|$.

Okay. Armed with these ideas about complex numbers, we can complete our desired task. Let s be any complex number. Then, in general, the control parameter and the position variable in

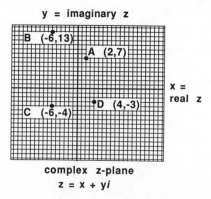

Figure 7.19   *Locating some points in the complex plane.*

equation (2) will also be complex. Let the complex position variable be written as $z_n = x_n + y_n \cdot i$ and the control parameter as $c = a + b \cdot i$. Equation (2) becomes

$$z_{n+1} = z_n^2 + c \qquad (7)$$

or, what is the same thing,

$$x_{n+1} + y_{n+1} \cdot i = (x_n^2 - y_n^2 + 2 \cdot x_n \cdot y_n \cdot i) + a + b \cdot i$$

Now, because the lefthand and righthand sides of the latter can be equal only if the real parts are equal *and* the imaginary parts are equal, we find that the equality of the real parts leads to equation (1a) and the equality of the imaginary parts leads to equation (1b). Equation (7) is really shorthand for equations (1a) and (1b). The Mandelbrot map is (a) a transformed, complex version of the logistic map, and (b) much more simply, equation (7). In this form, the dynamics space of the Mandelbrot map is the complex z-plane, whereas the control space is the complex c-plane. The Mandelbrot Set resides in the c-plane (its Julia Sets reside in the z-plane); it is the set of all complex c's for which the orbit of 0 (that is, $0 + 0 \cdot i$) is bounded under the map (equation 7). All the fantastic richness of the Mandelbrot Set emanates from the simple mantra "square z and add c, square z and add c, square z and add c,. . . . "

To reinforce the kinship of the logistic map and the Mandelbrot map, see Figure 7.20. Here the bifurcation diagram of the map $x_{n+1} = x_n^2 + a$, the transformed version of the logistic map, is shown in the upper portion. This bifurcation diagram is identical to that of the logistic map, except that it is flipped over because of the transformation given by equation (4). The leftmost edge of the diagram is at $a = -2$. The horizontal axis through the middle of the Mandelbrot Set below the bifurcation diagram is also the a-axis. The 1-cycle behavior of the bifurcation diagram sits over the main cardioid. The 2-cycle behavior sits over the head. The large 3-cycle window sits over a decoration on the spike that, under magnification, is revealed to be a midget (the "3-cycle midget"). In fact, every periodic window in the bifurcation diagram is associated with a midget.

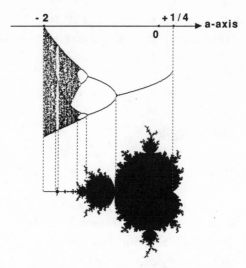

Figure 7.20   *A comparison of the bifurcation diagram of the map* $x_{n+1} =$ $x_n^2 + a$ *(upper), where a and x are real, with the Mandelbrot Set (lower). The bifurcation diagram is associated with the a-axis only. Note that the 1-cycle behavior in the bifurcation diagram coincides with the main cardioid in the M Set, the 2-cycle with the head, the 4-cycle with the bud on the head, and so on. Note also that windows of order are associated with large decorations along the spike. In particular, the 3-cycle window corresponds to a decoration that, when magnified, is a midget.*

Finally, you may be curious about what analogues of the Mandelbrot Set are produced by other simple, nonlinear maps. Figure 7.21 shows the sets of complex numbers, c, for which the orbits of 0 do not escape to infinity for the maps $z^3 + c$, $z^4 + c$, $z^5 + c$, and $z^6 + c$. As the exponent of z gets larger and larger, the generalized M Sets get more and more like a filled-in unit circle; the buds get smaller and the clefts get filled in.

A Gift to Be Simple

That so many people have spent so much time examining the Mandelbrot Set is due, in part, to the incredible diversity of forms evidenced by the Set at all levels of magnification. Surprises seem to emerge everywhere one looks. The Set's boundary

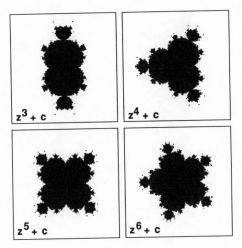

Figure 7.21   *Generalized "Mandelbrot Sets" for $z^3 + c$, $z^4 + c$, $z^5 + c$, and $z^6 + c$.*

is not strictly self-similar, though themes are certainly repeated again and again. The M Set is similar to an epic work of art. It is vast and has pleasing and subtle symmetry; variation and reinforcement are interwoven throughout it in the most delicate and unexpected balance. The visual and intellectual richness of the M Set emerges from a stunningly simple construction. The Mandelbrot Map that produces this graphic poetry—like its close cousin the logistic map—is a one-rule, iterated function system, but the rule, because it requires squaring, is not invertible. That only the slightest amount of nonlinearity could lead to such bizarre and wonderful complexity is the profound lesson of the Mandelbrot Set. Learning it shatters our preconceptions of what to call simple and what complex.

c h a p t e r

The Fragility of Order

ONE OF THE most interesting features of chaos is the unpredictability arising from sensitive dependence on initial conditions. Consider, for example, the logistic map with s = 4. We know that iterating two initially nearby states almost always gives sequences of states that become less and less alike as time goes on. Both states spawn chaotic time series, and the sensitivity to initial conditions amplifies the small difference between the starting values.

Unpredictability in endlessly unsettled, chaotic systems should no longer be a surprise. On the other hand, complete predictability for ordered systems with periodic motion seems a reasonable expectation. Unhappily, there is a sense in which this is not the case, either. In this chapter we shall see that in situations where there are several different ordered and stable outcomes of a nonlinear dynamical process, a tiny uncertainty in starting conditions can make it totally unpredictable which of the stable outcomes the system evolves into. This may be a more pernicious sort of unpredictability than that associated with chaos: when the ultimate behavior is orderly, we don't usually pay close attention the starting conditions, because we are accustomed to expect that small changes will have small effects in that case.

We begin with a few examples where the effect is fairly transparent, and we conclude with a warning story — an admonition to be wary of this manifestation of fractals in the world.

Basins of Attraction

You may have seen the toy shown in Figure 8.1. It consists of several magnets taped down on a flat plate with another magnet attached to the end of a rod and free to swing back and forth like a pendulum, but in any direction. Suppose the hanging magnet has its south pole pointing down and the fixed magnets have their north poles pointing up. If you have ever played with such a toy, you know that the motion that results when the pendulum is pulled aside and then let go is extraordinarily unpredictable; that's its charm. Eventually the erratic, zigzag motion of the pendulum slows enough for the hanging magnet to be attracted to one of the fixed magnets, near which it then comes to rest. This toy can be used to do an informative experiment. Here's how.

Paint the faces of three magnets black, gray, and white, respectively. Insert a sheet of paper between the plate and the three magnets, taping them to the sheet with their painted faces up. Pull the pendulum aside, and note where the center of the pendulum magnet starts off as a point projected vertically down onto the sheet (Figure 8.2). Let the pendulum go, and observe which of the three painted magnets it eventually gets stuck nearest. Put a small blob of black, gray, or white paint (corre-

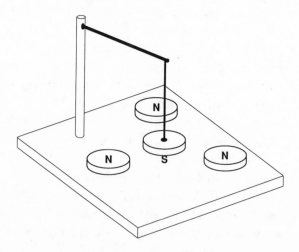

Figure 8.1 *A magnetic pendulum over three magnets.*

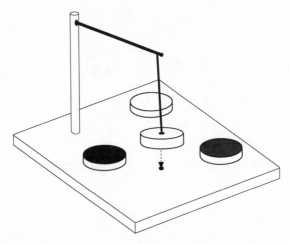

Figure 8.2 *Plotting a starting point for the magnetic pendulum experiment.*

sponding to the color of the sticking magnet) at the pendulum's
starting point. With enough patience, one eventually can color
the entire sheet. What does the resulting coloring look like?

A perfectly reasonable expectation, for an innocent in the
ways of nonlinear dynamics, is that the sheet will fill in something like what is shown in Figure 8.3. (In this figure, we're
looking down on the sheet from above.) Each of the uniformly
colored pie pieces represents the collection of all pendulum
starting points that are attracted to the magnet of the same color.
Such a collection has the evocative name *basin of attraction*. But,
you, attentive reader, surely suspect that nothing quite so simple
will happen. Indeed, what is actually observed is more like Figure
8.4. Though nearby starting points are often colored the same,

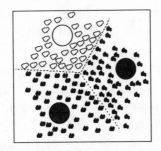

Figure 8.3 *A first guess at which starting points end up over which magnets.*

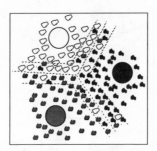

Figure 8.4   *Observation of where a few starting points end up over which magnets.*

sometimes there is an intermingling of colors. Instead of the reasonably sharp boundaries between the black, gray, and white basins shown in Figure 8.3, we find the actual boundaries to be diffuse. Some black and some white blobs appear in what might have been assumed to be a region of pure gray, and so on. This is not due to some sort of random fluctuation. If one could bring the pendulum back to exactly the same starting spot, it would produce the same result (though the word *exactly* is important here). The interlacing structure of the pendulum's basins of attraction is real.

Our little toy experiment is too crude to reveal the precise nature of the blurry basin boundaries. But if we could perform this experiment much more carefully, we would see a picture vaguely reminiscent of the last chapter, which dealt with the Mandelbrot Set. Near the boundary of the Mandelbrot Set is a pocked and pitted landscape. Black and colored patches intermingle in a confusion of peninsulas and fjords. The geometry of the boundary of the M Set is rewritten in its Julia Sets—in the collections of starting points in dynamics space for which the dynamics remains bounded. The boundary of a Julia Set separates basins of attraction: outside, infinity is the attractor; inside, one finds stable bounded behavior. Think again of the transporter metaphor of the last chapter. Imagine, for preset control values, carrying the transporter around to different launch points near the boundary of a Julia Set. If we set up within the set (in the black), the launch is safe (the orbit of the jumper is bounded); if we set up outside the set, the jumper is zapped off to infinity.

The task, near such a boundary region, of finding safe launch sites is equivalent to tip-toeing through a minefield of colossal intricacy. The room for error for a safe set-up is unimaginably small.

Another Transporter Machine

Our pendulum toy is a real example of a deterministic dynamical system with three attracting fixed points. The state values are the x- and y-coordinates of the center of the pendulum magnet at any instant, and the fixed points are the x- and y-coordinates of the center of the pendulum when it comes to rest near each of the three attracting magnets. The interleaving of the basins of attraction for these fixed points creates an interesting and complicated pattern. To study that pattern, and its consequences, for the pendulum toy would require a full description of all the forces acting on the pendulum (magnetic, gravitational, and frictional forces and stresses in the support mechanism) and an assessment of the resulting motions. Such a project is well beyond the scope of this book. Instead, we examine a two-dimensional map that, though we do not propose it as a faithful model of anything, contains the essence of the pendulum toy's behavior.

Like the Mandelbrot map of Chapter Seven, the map we wish to discuss here can be thought of as a transport machine in two dimensions. Technically called the Newton map for the cube roots of unit (whew!), it is defined by

$$x_{n+1} = \frac{2 \cdot x_n}{3} + \frac{x_n^2 - y_n^2}{D} \qquad \text{(1a)}$$

and

$$y_{n+1} = \frac{2 \cdot y_n}{3} - \frac{2 \cdot x_n \cdot y_n}{D} \qquad \text{(1b)}$$

where

$$D = 3 \cdot (x_n^2 + y_n^2)^2$$

Equations (1a) and (1b) zap a jumper from one place (x_n, y_n) after jump n to another $(x_{n+1}, y_n + 1)$ after jump $n + 1$. Unlike the Mandelbrot map, however, this transporter has no adjustable control knobs; it's always on and ready to go.

The Mandelbrot machine had all kinds of interesting jumping behaviors (orbits): escape to infinity, periodic attractors, and bounded chaotic attractors. The dynamics of the map defined here is less rich; it is dominated by three attracting fixed points. (That's why we said it's like the pendulum toy.) We can figure out where these attracting fixed points are from equations (1a) and (1b) by recognizing that at a fixed point, the jumper is stuck: $x_{n+1} = x_n = x_f$ and $y_{n+1} = y_n = y_f$. To find x_f and y_f, we express equations (1a) and (1b) as

$$x_f = \frac{2 \cdot x_f}{3} + \frac{x_f^2 - y_f^2}{D}$$

and

$$y_f = \frac{2 \cdot y_f}{3} - \frac{2 \cdot x_f \cdot y_f}{D}$$

where

$$D = 3 \cdot (x_f^2 + y_f^2)^2$$

Solving these equations for x_f and y_f is a bit of uninformative algebra; rather, we state the results and leave the steps as an exercise. We find that the fixed points are

$$(1, 0), \quad \left(-\frac{1}{2}, \frac{\sqrt{3}}{2}\right), \quad \text{and} \quad \left(-\frac{1}{2}, -\frac{\sqrt{3}}{2}\right)$$

These three fixed points are stable—that is, they are attractors. We won't show this in detail but will look at an example instead. Stability means that small perturbations die out. Equivalently, a fixed point is stable if all starting points sufficiently close to it "evolve back to it." Take $x_0 = 1.1$, $y_0 = 0$, for example. (Taking $y_0 = 0$ is convenient because in that case $y_n = 0$ for all n, according to equation (1b). This is a point near the fixed point (1, 0). A short table of subsequent jumps is

n	x_n	y_n
0	1.1	0
1	1.00881 . . .	0
2	1.0000768 . . .	0

Very quickly (x_n, y_n) goes back to (1, 0). That's just like the pendulum toy. When the pendulum is at rest over one of the three fixed magnets, a gentle tap will cause it to quiver, but soon the pendulum returns to rest where it started.

We want to chart basins of attraction, so we move the transporter around from one (x_0, y_0) to another and see what jumps result. We color the three fixed points (1, 0), $(-\frac{1}{2}, \frac{\sqrt{3}}{2})$, and $(-\frac{1}{2}, -\frac{\sqrt{3}}{2})$ black, white, and gray, respectively. We then color the starting point (x_0, y_0) the color of the fixed point to which the jumps terminate. Eventually, almost the whole (x, y)-plane will be painted black, white, or gray; the white (respectively, gray, black) region is the basin of attraction of the white (respectively, gray, black) fixed point.

Figure 8.5 shows a computer calculation of these basins and their intricate interleaving. Figure 8.6 shows the boxed portion of Figure 8.5 at higher magnification. The vaguely scalloped structure of Figure 8.5 repeats at many (indeed all) levels of magnification.

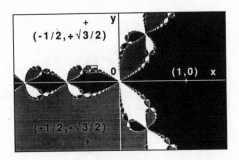

Figure 8.5   *The basins of attraction for the three fixed points of the Newton map for the cube roots of unity.*

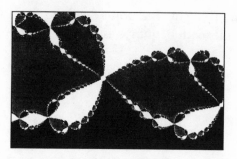

Figure 8.6   *A magnification of the box indicated in Figure 8.5.*

Figure 8.7 is a redrawing of Figure 8.5 except that all the coloring has been removed. What remains is the skeletal boundary between the basins of attraction. If one chooses launching points near the basin boundary, very slight differences can give rise to very different outcomes. Parenthetically, we note that if, by some ill fortune, a starting value exactly *on* the boundary were chosen, and if computations could be executed with infinite precision, successive iterations would fail to converge to any of the fixed points; rather, successive jumps would wander chaotically around the boundary forever. This chaotic motion is unstable, however, and any round-off errors will cause convergence to one of the fixed points (you just won't know which one).

As complicated as Figures 8.5, 8.6, and 8.7 are, they do not tell quite the whole story. They give no information about the dynamics. In Figure 8.8 we see an indication of how complicated the orbits can be. We might expect that the successive points generated by the map given as equations (1a) and (1b) would lie

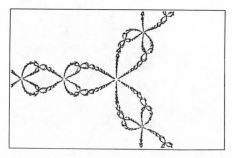

Figure 8.7   *The "basin boundaries" for the Newton map.*

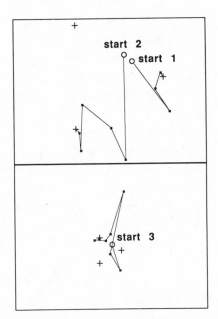

Figure 8.8   *Top: the orbits for the Newton map from two nearby starting points, start 1 and start 2. The fixed points are marked with crosses. Bottom: the orbit for the Newton map from a third starting point. Note how far the orbit moves from the three fixed points. The scale of this part is about four times larger than that of the top part.*

along a straight line connecting the initial point and the fixed point to which the jumps converge. In the top of Figure 8.8, we see that this need not happen at all. The bottom half of the figure shows something even more amazing: the jumps generated by our transporter machine can move very far away from all of the fixed points before finding their way back to one of them!

Newton's Method for Digging up Roots

Although it seems clear that the dynamics of equations (1a) and (1b) successfully captures the essence of the pendulum toy, you may well wonder how we arrived at the not-so-obvious form of the map. When we dealt with the same question for the Mandelbrot map in Chapter Seven, we pointed out (a) that the form emerged from a theoretical tie to the logistic map and (b) that the form was the (x, y) version of a map that was much simpler in appearance in complex number notation. Similar comments are true for equations (1a) and (1b) of this chapter. The story begins with "number crunching."

Frequently, in building mathematical models for describing natural or social phenomena, we are forced to deal with non-linear arithmetic that requires numerical solutions produced by computer. For example, the trajectory of a projectile fired near the surface of the Earth might be approximated by the relation $y = a \cdot x^2 + b \cdot x + c$, where y is the altitude of the projectile, x is its horizontal distance from some reference point, and a, b, and c are numbers describing launch conditions and gravity. Perhaps we would like to know where the projectile strikes the ground. We would need to find those special values of x for which y, the height above the ground, is 0 — that is, we would need to know the *roots* of the quadratic function y. Of course, the roots of a quadratic are easy to find: you can use the quadratic formula. Though there are similar (but extremely messy) formulas for the roots of cubic and quartic polynomials, a more general nonlinear relation has no analytic root extraction algorithm. Instead, we must resort to numerical recipes. Newton devised one of the most useful of these. It goes like this:

First, consider the problem of finding where the linear relation $y = m \cdot x + b$ has its root. Here m and b are taken to be known numbers. We now introduce some new notation, which may seem a bit obfuscatory at first but will prove useful. We say that y is the graph of a *function of x*, denoted by f(x). Of course, f(x) is just $mx + b$ for the case we are studying. We note that the *derivative* of f(x) (a concept we introduced in Chapter Five), designated by $f'(x)$, is just the slope, m. (Why? By definition, $f'(x) = [f(x + d) - f(x)]/d$ in the limit as d goes to 0. But $f(x + d) = m \cdot x + m \cdot d + b$, so $f(x + d) - f(x) = m \cdot d$. Thus $f'(x) = m \cdot d/d = m$.) Okay. Now Newton's method is to pick any point x_0 as a guess for where f(x) = 0. Evaluate $f(x_0)$; unless you were very lucky, it won't be zero. So draw a line through the point $(x_0, f(x_0))$ with slope equal to $f'(x_0)$ (=m). See Figure 8.9. There are

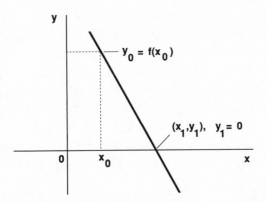

Figure 8.9   *Newton's method for finding the root of a linear equation.*

many lines with slope $f'(x_0) = m$, but the line drawn is the *correct* line, $y = m \cdot x + b$, because it has the right slope and because one of its points — $(x_0, f(x_0))$ — is known to be on the correct line. Where the line so constructed crosses the x-axis is the point $(x_1, 0)$; x_1 is the desired root. The numerical value of x_1 is easy to determine. It comes from calculating the slope:

$$f'(x_0) = \frac{f(x_1) - f(x_0)}{x_1 - x_0} = \frac{0 - f(x_0)}{x_1 - x_0}$$

or, after rearranging,

$$x_1 = x_0 - \frac{f(x_0)}{f'(x_0)} \qquad (2)$$

Let's try an example. Suppose $f(x) = -13x + 6$. What is the root, x_1, of f? This is an extremely simple problem. Set $f(x_1)$ equal to 0 and solve for x_1: $x_1 = \frac{6}{13}$. Let's see how Newton's method produces the same result. Pick any guess: $x_0 = 5$, say. Now, $f'(x_0) = -13$. Thus, according to equation (2),

$$x_1 = 5 - \frac{-13 \cdot 5 + 6}{-13} = 5 - \frac{59}{13} = \frac{6}{13}$$

as expected.

Nobody uses Newton's method to solve linear problems. Its real value is in finding nonlinear roots. Let y be the graph of f(x), where f is not linear. The procedure starts off as described above, except that now, through the point $(x_0, f(x_0))$ we draw the tangent line to the curve $y = f(x)$. This tangent line has the slope $f'(x_0)$, where f' is the derivative of f, defined on page 286. The tangent line so drawn intersects the x-axis at a point x_1 that now is *not* the correct answer because $y = f(x)$ curves away from its tangent line (otherwise, we would be doing the linear problem again). See Figure 8.10.

Here is Newton's neat trick: Use x_1 as a new guess. Calculate $f(x_1)$. Through $(x_1, f(x_1))$ construct a new tangent line with slope $f'(x_1)$; the new tangent line intersects the x-axis at a point x_2 that also is *not* the correct root. But if x_0 wasn't too poor an initial guess, x_2 will be closer to the correct answer, x_r, than was x_1. Obviously, we are encouraged to repeat the process again and again, expecting ultimately to get arbitrarily close to the root x_r. In short, Newton's method is an iterative process in which successive guesses at the correct answer are generated from previous ones through a generalization of equation (2):

$$x_{n+1} = x_n - \frac{f(x_n)}{f'(x_n)} \tag{3}$$

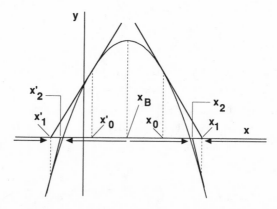

Figure 8.10 *Newton's method applied to a nonlinear equation. Note how different first guesses can lead to different roots.*

Let's try a nonlinear example. Suppose $y = -5 \cdot x^2 + 6 \cdot x + 2$. The roots of this quadratic equation are again easy to find using available analytic tools—in this case, the quadratic formula.

$$x_r = \frac{-6 \pm \sqrt{6^2 - 4 \cdot (-5) \cdot (2)}}{-10} = 1.471 \ldots \text{ and } -0.271 \ldots$$

(Recall that a first-order polynomial has one root, a second-order polynomial has two roots, a third-order polynomial three, and so on). Now let's use Newton's method. We will need a general expression for the slope of the tangent line to f(x), because like f, f′ has to be evaluated at a number of different x's. To find the general form for f′(x), we employ the rule developed in Chapter Five that the derivative of f(x) is the limit of the ratio [f(x + d) − f(x)]/d as d goes to zero. In this case,

$$f(x) = -5 \cdot x^2 + 6x + 2$$

so

$$\begin{aligned}
f(x + d) &= -5 \cdot (x + d)^2 + 6 \cdot (x + d) + 2 \\
&= -5 \cdot (x^2 + 2 \cdot x \cdot d + d^2) + 6 \cdot (x + d) + 2 \\
&= -5 \cdot x^2 + 6 \cdot x + 2 - 10 \cdot x \cdot d + 6 \cdot d + 5 \cdot d^2 \\
&= f(x) - 10 \cdot x \cdot d + 6 \cdot d - 5 \cdot d^2
\end{aligned}$$

Thus,

$$\begin{aligned}
\frac{f(x + d) - f(x)}{d} &= \frac{-10 \cdot x \cdot d + 6 \cdot d - 5 \cdot d^2}{d} \\
&= -10 \cdot x + 6 - 5 \cdot d
\end{aligned}$$

In the limit as d goes to zero, the latter becomes just

$$f'(x) = -10 \cdot x + 6$$

Let's try $x_0 = 5$. We tabulate successive values of x_n using equation (3).

n	x_n
0	5
1	2.886 . . .
2	1.909 . . .
3	1.544 . . .
4	1.474 . . .
5	1.471 . . .

In five iterations we have achieved at least four-place agreement between the analytic and approximate results for one of the roots.

Suppose that instead of $x_0 = 5$, we had chosen $x_0 = 0$.

n	x_n
0	0
1	-0.333 . . .
2	-0.273 . . .
3	-0.271 . . .

Here we have four-place agreement to the second root in only three iterations. Clearly, the choice of x_0 is very important in determining to what root Newton's method converges. In Figure 8.10 we see a plot of a quadratic function like the one of this example. All x_0's to the right of x_B converge to the larger root; all x_0's to left of x_B converge to the smaller. (*Note:* x_B is where y takes its maximum value. Where is that for this example?)

If we think of the iteration scheme of Newton's method (embodied by equation 3) as a dynamical system, its attractors are the roots of the function f(x). Each root has its own basin of attraction.

For the quadratic functions in the preceding example, the basin boundaries are simple points (such as x_B in Figure 8.10).

Successful implementation of Newton's method for these problems requires finding these simple boundary points and making sure that x_0 is chosen in the appropriate basin. Where f is more complicated, however, the basin boundaries are often fractal-like. It's not so easy to make sure you've chosen a starting guess in the proper basin in these cases; it's similar to the task of tip-toeing around the fractured boundary of a Julia Set.

To see what the problem is, let's consider a concrete example of root finding in two dimensions—namely, finding the roots of the complex function

$$f(z) = z^3 - 1$$

The values of z for which this $f(z) = 0$ are called the *cube roots of unity*. The history of this problem is interesting. In 1879 Arthur Cayley [112] applied Newton's method to solving the equation $z^2 = 1$ (that is, to finding the values of z for which $z^2 - 1$ is zero). Of course, we don't need Newton's method to solve this equation: the solutions are just $z_r = 1$ and $z_r = -1$. What Cayley did was to study which initial guesses lead to sequences converging to $z_r = 1$ and which initial guesses lead to -1. Perhaps not surprisingly, he found that any complex number with its real part positive [that is, $z = x + y \cdot i$ with $x > 0$, corresponding to the point (x, y) with $x > 0$] converges to $+1$ and that any complex number with its real part negative converges to -1. Those are the two basins of attraction (though Cayley didn't call them that). He also observed that if the initial guess has real part zero, then the sequence doesn't converge to either $+1$ or -1. In fact, it doesn't converge to anything at all but wanders forever on the imaginary axis—the vertical line through the origin. Subsequently, Cayley announced his intention to study the same problem for $z^3 = 1$. He died 16 years later, frustrated in these efforts. The solution was finally obtained in 1984. Here's why it took so long.

Because $f(z) = z^3 - 1$ is a cubic, we know there will be three roots. In fact, they are also easy to extract without resorting to Newton's method. One of them is obviously $z_r = 1$. For the roots that do not equal 1, we can divide $z^3 - 1$ by $z - 1$ (that is not 0 as long as z is not 1, and we're not interested in z = 1 because we already know it is a root). This division produces the quadratic $z^2 + z + 1$. We obtain the roots of the quadratic by using the quadratic formula. After doing so, we find that the three roots of $z^3 - 1$ are

$$z = 1, \quad \frac{-1 + \sqrt{3} \cdot i}{2}, \quad \frac{-1 - \sqrt{3} \cdot i}{2}$$

Note that the latter two roots are complex; this means that Newton's method for finding the roots of $z^3 - 1$ is a (one-rule) nonlinear iterated function system operating in the complex plane. In that sense it is similar to the Mandelbrot map. You can expect fractals popping up everywhere.

To use Newton's method to obtain the three roots of $f(z) = z^3 - 1$, we need the derivative of $f(z)$:

$$\frac{f(z + d) - f(z)}{d} = \frac{[(z + d)^3 - 1] - (z^3 - 1)}{d}$$

$$= \frac{z^3 + 3 \cdot z^2 \cdot d + 3 \cdot z \cdot d^2 + d^3 - 1 - z^3 + 1}{d}$$

$$= 3 \cdot z^2 + 3 \cdot z \cdot d + d^2$$

[To get the next to the last equality, we used the fact that $(z + d)^3 = z^3 + 3 \cdot z^2 \cdot d + 3 \cdot z \cdot d^2 + d^3$.] Allowing d to go to zero leads to

$$f'(z) = 3 \cdot z^2$$

Thus Newton's method for $f(z) = z^3 - 1$ becomes

$$z_{n+1} = z_n - \frac{z_n^3 - 1}{3z_n^2} \tag{4}$$

Because z_n is complex, this relation is really one rule for changing x_n (the real part of z_n) in each iteration and a second for changing y_n (the imaginary part). Like the Mandelbrot map, it's a "machine" for transporting points around in the plane.

Expressed in real and imaginary parts, Newton's method for finding the roots of $z^3 - 1$ is exactly equivalent to the set of transformations given in equations (1a) and (1b). As we noted, the basin boundaries for equations (1a) and (1b) are amazingly complicated, so determining which initial guesses z_0 converge to which roots is a beastly problem. On one level, the lacy trinket revealed in Figure 8.7 — the basin boundary — appeals to us as an object of visual richness. On another, the intricacy of the boundary underscores that the problem of finding the roots of a nonlinear function by numerical means is extremely subtle. No wonder that Cayley, armed only with paper and pencil, couldn't figure out what was going on!

Another Surprise in Newton's Method

Newton's method has yet another surprise in store for us. In 1983 J. Curry, L. Garnett, and D. Sullivan [113] studied Newton's method for the family of cubic functions

$$F_c(z) = z^3 + (c - 1) \cdot z - c$$

where c is a complex number. Note that whatever value is taken by the complex number c, $z = 1$ is a solution of the equation

$$F_c(z) = 0$$

Curry, Garnett, and Sullivan let c vary over some region of the complex plane. In Figure 8.11 the region shown has both the real and the imaginary parts of c lying between -2 and 2. For the same reason that the Mandelbrot iteration begins with $z = 0$, for each c they applied Newton's method starting with $z = 0$. If the sequence that was generated converged to $z = 1$, the pixel corresponding to c was painted black; otherwise, it was painted white. Figure 8.11 is the result of this experiment. It looks fractal, but not all that different from other pictures we have seen in our examination of Newton's method. What's so interesting about this picture?

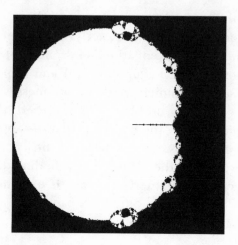

Figure 8.11   *Newton's method for $F_c(z)$ as c varies over the complex plane.*
The starting guess is always z = 0. The pixel corresponding to c is painted black
if Newton's method converges to the solution z = 1; otherwise, it is painted white.

As with the Mandelbrot iteration, the Newton sequence can
do other things than converge to a fixed point. We have already
mentioned that if by bad luck we started the iteration at a point
on the boundary of a basin of attraction, then Newton's method
would wander chaotically forever, never converging to anything.
This sort of problem we can ignore, because the basin boundaries
are very sparse and hence difficult to find numerically. We shall
see that more serious problems can occur, however.

Curry, Garnett, and Sullivan modified their experiment by
adding another color: they painted c gray if Newton's method
converged to some cycle but not to a fixed point. The left half of
Figure 8.12 shows a magnification of the top "bug" of Figure 8.11, with this new gray shading added. The right half
of the figure magnifies the gray region, showing clearly what it is.
The Mandelbrot Set has arisen from a problem unrelated to the
original iteration of $z^2 + c$. Unlike the familiar Mandelbrot Set,
for c in the main cardioid of *this* copy, the Newton sequence
converges to a 2-cycle; for c in the head, the Newton sequence
converges to a 4-cycle; and in general, the periods are twice
those of the familiar Mandelbrot Set. Of course, an infinite cloud
of Mandelbrot midgets surrounds this set, and the other bugs of
Figure 8.11 have their own Mandelbrot Sets of c leading to

 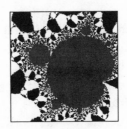

Figure 8.12   *Left: a magnification of the "bug" near the top of Figure 8.11, with the modification that pixels are painted gray if Newton's method does not converge at all. Right: magnifying the gray region reveals a familiar shape.*

periodic points, though the periods are other multiples of the usual periods.

These unexpected Mandelbrot Sets are important, because they measure the regions on which Newton's method fails to converge to any answer. What is also becoming increasingly clear is that the Mandelbrot Set is much more common than had been imagined at first. A lot of reasonably complicated families of functions have small copies of the Mandelbrot Set lurking around in them, so the time we spent studying the Mandelbrot Set was not simply time devoted to a peculiar — if pretty — mathematical curiosity.

A Parable

As far as we can tell, the universe is an intrinsically nonlinear place. Linear behavior, wherever it seems to surface, only approximates more general phenomena. Nonlinearity is the source of the diversity and apparent complexity surrounding us. Schooled in the Newtonian paradigm of linearization, we are constantly challenged to recall that the real, nonlinear world is filled with nonlinear peculiarities. Sticking to a linear path in a nonlinear terrain can inflict painful lessons.

It is instructive to consider the following example. This example is not intended to be realistic in detail, but its point is real enough. Let's consider the dilemma of operating a nuclear reactor. A reactor is a complex physical system whose operation depends on some details of nuclear physics, thermodynamics, and fluid mechanics. It is certainly a nonlinear device. The operation of the reactor is designed to convert nuclear energy into heat, which then might be used to generate electricity. The reactor runs in a "critical" state: one nuclear event produces exactly one subsequent nuclear event. If the reactor is running subcritically, it produces heat too slowly to be an effective generator of electrical power. (This is similar to simple feedback with losses outstripping gains — that is, with a gain less than 1.) If the reactor runs supercritically, the nuclear processes release energy faster and faster (feedback with a gain greater than 1); the reactor overheats, and all manner of unpleasantness ensues. The latter situation is referred to as "melt down."

Constructing a detailed mathematical model for a real reactor would require substantially more formal apparatus than we care to get into here. For our story, it is sufficient to treat our reactor in a qualitatively suggestive way. Thus we consider the reactor to be described by three possible states: "shut down," "safe," and "melt down." Further, we compress the elaborate control of the reactor into a single, fictitious tuning knob, the setting of which establishes the operating parameters. We imagine, then, a console in the reactor's control room with a solitary dial like the one shown in Figure 8.13. In a linear world, crisp linear boundaries would separate dial settings for the three operating states. The setting shown in the figure might then be thought of as a kind of conservative operating strategy: running comfortably in the safe zone but with control near shut down so that any unexpected fluctuation could be handled without danger.

Figure 8.13 *A simple cartoon of the control of a nuclear reactor.*

Although the face of our imagined dial may well be marked off in clean pie pieces as shown, the reactor state is actually driven by nonlinear dynamics: underlying the face of our dial there is probably a space of state values housing fractal monsters at every turn. Our little story is nothing less than a tale of horror. Let us suppose our reactor works as follows: The tip of the dial picks a point in the plane within a disk of radius 1 centered at the origin. This point sets off a chain of physical events described mathematically by a sequence of new values leading to one of three attracting states; suppose the three attractors are taken to be the three fixed points of equations (1a) and (1b). Oh sure, no reactor is going to behave exactly like this cartoon, but the qualitative aspects of our description are not far off the mark. Disguised by the face of the reactor's controlling dial is a state space interlaced with fractal basin boundaries. This disguise is unmasked in Figure 8.14. There we see the stunning truth: our conservative strategy keeps the reactor teetering on the precipice of tragedy. A little fluctuation in the wrong direction and it's a melt down. Not only that, but Figure 8.14 also suggests that sometimes the obvious plan for returning to safety can be disastrous. Cranking the dial from a safe position directly to "shut down" through the piece of the basin for melt down that stands in the way can produce very undesirable results.

If you think this story goes too far to make its point, please consider the case of the accident at the Chernobyl electric power reactor. On April 26, 1986, at a reactor facility about 60 miles north of the city of Kiev in the former Soviet Union, a series of

Figure 8.14 *A reactor control revealing more complicated settings. Note that moving clockwise from "safe" toward "shut down" traverses a large "melt down" region.*

catastrophic events led to a melt down. The concatenation of unlikely occurrences that culminated in this deadly incident (29 people died as only an *immediate* consequence) was punctuated with a disastrous piece of nonlinearity. It is on the latter that we wish to focus our attention.

To appreciate the problem, you need to know a few facts about the nuclear processes involved in a fission reactor. The fuel of such a reactor is slightly enriched uranium. In its natural state, uranium comes in essentially two forms, or isotopes: uranium-235, the fissionable isotope (about 0.7%), and uranium-238, the nonfissionable isotope (about 99.3%). When a nucleus of uranium-235 absorbs a neutron (an electrically neutral, nuclear constituent), it divides (fissions) into lighter nuclear pieces, releasing energy and neutrons in the process. These neutrons can then be used to cause other uranium-235 nuclei to fission, producing what is known as a chain reaction. The neutrons that are released in fission initially travel very fast. Uranium-235 absorbs slow neutrons well, fast neutrons poorly. Uranium-238, on the other hand, absorbs fast neutrons well and, in so doing, is eventually transformed to another nuclear species, plutonium-239, but without the release of energy. To create a chain reaction, two things are done: uranium-235 is made more plentiful (up to a few percent) in the fuel elements of the reactor via chemical means, and other nuclei, which do not absorb neutrons, are inserted into the fuel elements to slow the fission-product neutrons down. The latter material is called *moderator*. If the reactor is *under*moderated, it can't sustain a chain reaction because the fission neutrons are too fast to be absorbed by the fuel isotope of uranium. If the reactor is *over*moderated, the distance that a slow neutron has to travel before possibly being absorbed by a uranium-235 nucleus is too great, and competing events (loss through the walls of the fuel rods or absorption by some other nuclear species) mitigate further fissions. Thus a reactor that is capable of sustaining a chain reaction has only a narrow range of allowed moderation. Figure 8.15 shows this: only between moderator values A and B can a chain reaction occur.

A properly functioning, controlled reactor flirts continuously with disaster. If less than one fission neutron produces a second fission for any length of time, the reaction chain breaks and the

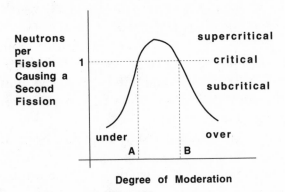

Figure 8.15   *The criticality of a fission reactor versus the amount of moderator material in the fuel elements.*

reactor shuts down. If more than one fission neutron produces a second fission, then the reaction runs away and melt down occurs. A controlled chain reaction requires one secondary fission for every primary fission. Consequently, only the moderator values A and B shown in Figure 8.15 lead to a controlled chain reaction. The engineering of the reactor design strives to ensure maintenance of such values. Actually, the undermoderated critical value A is preferred, because a sudden loss of moderator there produces an automatic shut down. A sudden loss of moderator while the reactor is operating at B can be calamitous: unless rectified immediately, it can cause melt down.

Back to Chernobyl. In the reactor at Chernobyl, water was used to carry away the heat produced by fission in the fuel element core. Collisions between the fission-product neutrons and the nuclei of the cooling water actually help moderate a fission reactor. A long sequence of unusual events on the day of the tragedy had placed the Chernobyl reactor in an overmoderated state (like state B in Figure 8.15). Unfortunately, while it was in that state, a temperature excursion caused the cooling water to begin to boil, and a depletion of moderator ensued in the core. Faithfully adhering to the nonlinearity of the curve depicted in Figure 8.15, as the degree of moderation moved from state B toward state A, the reactor raced into supercriticality, and in a matter of seconds the core melted, a chemical explosion occurred, and a flash fire was ignited.

Although the lessons of Chernobyl are well worth studying, our purpose was not to engage in a polemic about the dangers of the nuclear power industry. We merely wish to emphasize the point that in the nonlinear universe, existence near a fractal basin boundary can be irreversible and unforgiving. The most innocent decisions, the most innocuous choices, can produce the most profound outcomes.

On a less urgent plane, if only we could see the fractal landscape through which our life lines pass, how wisely we could control our destinies. But that would also take a lot of the fun out of it. Isn't it more exciting not knowing which great attractor will pull us this way and which that?

c h a p t e r

Some Call It Life

WE BEGAN THIS survey of complexity in Chapter One with a description of video feedback. In video feedback, a TV monitor is connected to the output of a TV camera, which in turn receives its input from the light emitted by the monitor's screen. When tuned properly, this optoelectronic loop generates light patterns that have both space complexity and time complexity. A fruitful way of thinking about the light on the screen is to visualize the screen as a field of pixels. Each pixel is a little cell that can have any one of several color states (including black, or off). The states of all pixels on the screen are changed in discrete time steps as the video display is refreshed every few milliseconds. In the next screen update, the state of a given pixel (the light it emits) depends on its current state plus the states of all other pixels in the field through some, perhaps complicated, set of deterministic rules. Thought of this way, video feedback is an example of what is known as a cellular automaton—a collection of cells, spread out in space, whose states are changed at regular intervals in time in a machinelike manner according to deterministic rules. Even simple automata can exhibit remarkable properties, some of which are quite close to what we observe in living organisms. For this reason, there is a keen interest in discovering what cellular automata can teach us about how we came to be, how we think, and how we behave.

Cellular Automata

In the late 1940s the mathematician John von Neumann became interested in "synthesizing" living creatures. As an important step in this program, he studied artificial self-reproducing systems. His initial conceptual designs were mechanical: robot islands that would extract energy and raw materials from the lake in which they floated and build copies of themselves. In physical implementation, these machines would have been extremely complex, consisting of hundreds of thousands of parts. It is unlikely that von Neumann actually entertained thoughts of seeing them constructed. His real goal was to understand the principles of self-reproduction, yet much of the intricacy of his designs dealt with peripheral problems — for example, getting the robot to identify raw materials moving in the lake and grasp them with its mechanical hands. The difficulty of this part of the problem is underscored by observing that even today, after a tremendous amount of work by many clever people, robot vision and manipulation of the environment are far from perfected.

Eventually Stanislaw Ulam, who had earlier used the Manhattan Project computers to gain great insight into producing recursive geometrical patterns, suggested that the true character of self-reproduction might be more easily studied in an abstract geometrical space than in the three-dimensional world of physical robotics. This suggestion led von Neumann to an important conceptual breakthrough. Automata theory was born [114].

A *cellular automaton* is determined by four features:

The state space
The number of states per cell
The neighborhood of a cell
The rule

The *state space* is the collection of all cells (assumed identical) on which the automaton works. If the space is one-dimensional, then the cells are visualized as little intervals along a line; if the space is two-dimensional, the cells are usually pictured as squares, equilateral triangles, or regular hexagons.

The second aspect of a cellular automaton is the specification of the *number of states per cell*. Binary automata — those that have two states per cell — are the easiest to deal with and have been investigated the most widely, but more complicated choices are useful in some situations. For example, von Neumann invented a self-reproducing automaton with 29 states per cell. (We shall describe an interesting and practically important multistate automaton a little later.) For binary automata, the two states are often said to be "on" and "off" or "occupied" and "unoccupied" or "alive" and "dead"; a numerical representation of the states might be 1 and 0.

The third aspect of a cellular automaton is the specification of what constitutes the *neighborhood* of a cell — that is, the cells in the vicinity of a given cell that can influence the state of that given cell in the next generation. The neighborhood may include the given cell itself. For one-dimensional automata, the most common choices for the neighborhood of a cell are one or two cells on either side of the given cell. For the two-dimensional case with square cells, two particular choices of neighborhood have been studied extensively: a *von Neumann neighborhood* consists of the four cells north, south, east, and west of the given cell, and a *Moore neighborhood* consists of the eight cells nearest the given cell — that is, the von Neumann neighborhood plus the cells at the northeast, northwest, southwest, and southeast corners. In Figure 9.1, these neighborhoods (of the cell with vertical stripes) are highlighted with diagonal stripes.

The final aspect of a cellular automaton is *an evolutionary rule* that determines the state of a cell in the next generation from the states of the cells in its neighborhood in the present generation. By the configuration of a neighborhood we mean the arrangement of possible states in the cells of the neighborhood. For binary automata, the rule may be thought of as selecting, from

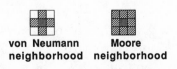

von Neumann **Moore**
neighborhood **neighborhood**

Figure 9.1 *Von Neumann and Moore neighborhoods.*

among all possible neighborhood configurations, those that result in the cell's being occupied in the next generation. For more general, multistate automata, the rule selects which neighborhood configurations result in the cell's being in the first state, which configurations result in the cell's being in the second state, and so on. An automaton evolves in fixed time steps. That is, we give some initial specification of the state of each cell — the "first generation" of cells. Then, for each cell in the first generation, the neighborhood is analyzed and the rule applied. This determines the state of each cell in the "second generation." In an actual calculation, two copies of the cell space are required: one containing the old generation and a second, initially empty one that is filled in by referring to the old copy. This method of producing the new generation is called synchronous updating, because in principle, each cell in the new generation could be turned on or off synchronously — at the same time. The process is repeated to produce successive generations. Clearly, cellular automata are dynamical systems; future states of cellular automata are completely determined by the states of the immediate past.

Before looking at some examples, we consider an issue arising from the difference between theoretical models and actual computer programs. Ideally, the state space would be infinite in extent, but of course any realization in a computer can contain only a finite number of cells. Perhaps the most common way of treating this situation is by imposing a "wraparound." In the one-dimensional case, the left-most cell and the right-most cell are taken to be adjacent. In the two-dimensional square case, the left-most and right-most cells in each horizontal row are adjacent, and the top and bottom cells in each vertical column are adjacent. There are other possibilities, though. For example, all the cells around the edge of the square array might be kept constantly in a particular state. The impact of the boundary on the evolution of the automaton is subtle, and we shall be content to consider only the wraparound case. As an illustration, Figure 9.2 shows the 300th generation of the pattern evolving from a single on-cell by a Moore automaton on a 100×100 grid (left) and the 297th generation on a 99×99 grid (right), both with wraparound. (Both represent three complete wraparounds of

Figure 9.2 *Two cell patterns emerging from identical rules and the same starting condition. The left pattern resides on a 100 × 100 grid, the right on a 99 × 99 grid. Each picture is taken after three complete wraparounds of the respective screen.*

the screen; the different numbers of generations follow from the different grid sizes. But note that the 300th generation of the 99 × 99 grid looks different from the 300th generation of the 100 × 100 grid, too.) Despite these pictures' being generated by the same automaton rule and from the same starting configuration, they bear little resemblance to one another. The boundary of the state space can have a profound effect on the evolution of the pattern.

To see how cellular automata behave, we'll work through a few one-dimensional examples. The state space can be viewed as a long string of boxes: ⊞⊞⊞⊞⊞⊞⊞⊞⊞ . Let's suppose that our cellular automaton is binary — that is, each cell is in one of two states. The states are depicted by ■ for "on" and by □ for "off." (Remember, you can insert "alive" for "on" and "dead" for "off.") The neighborhood consists of nearest neighbors only. In our first example, we choose the rule "If the central cell is on in a given generation but neither of its nearest neighbors is on in that generation, then it remains on in the next generation; otherwise, it goes off." Graphically, this rule is represented by □■□. Now in order to proceed, we need a starting configuration. Suppose we have one. The next generation is produced by examining the starting configuration for three-cell arrangements that look like the rule □■□ . Wherever such an arrangement occurs, the middle cell is turned on (in the case we are studying, the middle cell is *left* on). All other cells either remain off or are turned off.

With this rule, a starting configuration in which all cells are off remains that way because there are no "off-on-off" arrangements anywhere. On the other hand, suppose the initial configuration is ⊞⊞⊞⊞⊞■⊞⊞⊞⊞⊞ — one cell on somewhere in the space of cells. In this configuration there is exactly one ☐■☐ arrangement. In the next generation, because of the rule we are using, all cells are off except the one that was originally on. Successive generations would look like Figure 9.3a. Now suppose the initial configuration consists of two contiguous on-cells: ⊞⊞⊞⊞⊞■■⊞⊞⊞⊞⊞. Neither on-cell has *both* nearest neighbors off, so each will go off in the next generation, leading to a completely off configuration (which we know remains completely off subsequently).

A more complicated initial configuration would consist of some isolated on-cells (on-cells with no neighbors on) and some contiguous on-cell strings (that is, on-cells with at least one neighbor also on). All contiguous arrangements will go off in the next generation, whereas isolated on-cells will remain on. Thus, as an example, we would have Figure 9.3b.

So much for the rule ☐■☐. Suppose that instead we select the rule ☐☐■. This rule says, "If an off-cell (the central one) has *only* its right neighbor on, then that cell will be on in the next generation; otherwise, it will remain off. How does this change things? Start with the single on-cell configuration ⊞⊞⊞⊞⊞■⊞⊞⊞⊞⊞ . There is exactly one ☐☐■ arrangement in this configuration. The middle cell of that arrangement will be on in the next generation; all others will be off. The net effect is that in each successive generation, the cell immediately to the

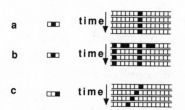

Figure 9.3 *The first few generations for different starting configurations under the rules* ☐■☐, *(a and b) and* ☐☐■ *(c).*

left of the cell that is currently on turns on (while the cell that is currently on turns off), as shown in Figure 9.3c.

As a final example, we take the rule "If a cell that is currently off has one and only one on-neighbor (either right or left), then it turns on; otherwise, it is off." Symbolically, this rule can be represented by ▪☐☐ or ☐☐▪. How does the single on-cell configuration ☐☐☐☐☐☐☐▪☐☐☐☐☐☐ evolve? Well, the first part of the rule by itself would cause the single on-cell to effectively shift one position to the right in each new generation. On the other hand, the second part of the rule by itself would cause the on-cell to effectively shift one position to the left in each new generation. Taken together, these parts allow on-cells to *multiply*. It happens as follows: If the first-generation transformation looks like ☐☐☐☐☐☐▪☐☐☐☐☐☐, the second is ☐☐☐☐☐▪☐▪☐☐☐☐☐ and the next ☐☐☐☐▪☐☐☐▪☐☐☐☐. (Note that in the latter, we don't get an on-cell directly in the middle because the neighborhood configuration ▪☐▪ is not one of the two specified in the rule.) Subsequently (as you should be able to check), come ☐☐☐▪☐▪☐▪☐▪☐☐☐, ☐☐▪☐☐☐☐☐☐☐☐▪☐, ☐▪☐▪☐☐☐☐☐▪☐▪☐, and so on. It's amusing and instructive to run this evolution on for more generations. Figure 9.4 shows such a run where the cell rows have been shrunk to a single pixel in height, and successive generations are crammed close together to save space. The time-lapse photograph that Figure 9.4 is looks remarkably like the Sierpinski Gasket. (Of course, it isn't exactly the Gasket, because a sufficient magnification of a piece of Figure 9.4 would not just show the same thing but would produce isolated pixels similar to

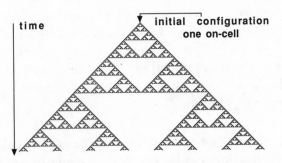

Figure 9.4 *Succeeding generations of on-cells stemming from a single on-cell under the rule* ▪☐☐ *or* ☐☐▪ .

the ones we have been discussing.) Despite its simplicity, the cellular automaton with the rule ■□□ *or* □□■ can lead to unexpectedly intricate patterns in space and time. That cellular automata can develop spatial and temporal structures suggests their potential utility in helping us understand fluid flows, aggregation phenomena (such as DLA), phase transitions (liquids becoming solids), morphogenesis, and population dynamics. Cellular automata models for such phenomena are currently under active investigation.

How Many Automata Are There?

Cellular automata are so simple that you might have trouble believing many interesting things could happen. One way to convince yourself that automata present some real possibilities, though, is to count the number of automata of a given type.

For example, consider one-dimensional, binary automata with the neighborhood of a cell consisting of the cell itself and one cell on each side. How many arrangements of on-cells and off-cells are possible for such a neighborhood? The left cell can be on or off: two possibilities. Regardless of the state of the left cell, the central cell can be on or off, and regardless of the states of the left and central cells, the right cell can be on or off. Thus the possibilities for (left   central   right) are (000), (001), (010), (011), (100), (101), (110), and (111). (Here 1 represents an on-cell, 0 an off-cell.) That is, there are $8 = 2^3$ arrangements of on and off cells for this type of neighborhood. By the same argument, for binary automata with the neighborhood of a cell consisting of the central cell and *two* cells on both the left and right sides, there are $2^5 = 32$ arrangements. In general, for automata with S states for each cell (S = 2 for those considered so far) and neighborhoods consisting of N cells, there are S^N arrangements of the neighborhood.

Returning to the S = 2, N = 3 example, each of the $8 = 2^3$ neighborhood arrangements can lead to the central cell's being in either state. How many rules can we construct from these possibilities? A rule is a sentence with, in this example, 8 clauses:

"If (1) (000) then the central cell (∗∗∗), or (2) (001) then the central cell (∗∗∗), or . . . , or (8) (111) then the central cell (∗∗∗)."

The (∗∗∗) in this sentence stands for one of the following two possibilities: "is on in the next generation" or "is off in the next generation." For example, the rule ■□ or □■, which can be expressed "If (100) or (001) then the central cell is on (in the next generation); otherwise, it is off," really means "If (000) then the central cell is off, if (100) then the central cell is on, if (010) then the central cell is off, if (001) then the central cell is on, if (110) then the central cell is off, if . . . , if (111) then the central cell is off." Because there are eight independent clauses, each with one of two possible endings, we can construct $2 \cdot 2 \cdot 2 \cdot 2 \cdot 2 \cdot 2 \cdot 2 \cdot 2 = 2^8 = 2^{2^3} = 256$ such rules. In general, if the central cell can be in one of S states, and there are S^N neighborhood arrangements, then a rule is a sentence with S^N clauses, each with one of S endings, and there are a total of S^{S^N} different such sentences.

We have seen that for $S = 2$ and $N = 3$, there are 256 automata. If you spent 1 minute examining each automaton, you would investigate all the possibilities in about $4\frac{1}{4}$ hours: an afternoon's work.

For $S = 2$ and $N = 5$ there are $2^{25} = 4,294,967,296$ automata. Let's suppose your practice with the $N = 3$ case improved your speed so much that you can examine an automaton in a second. To look at all these possibilities, you would have to work every second of every minute of every hour of every day of every week of every month for over 136 years. There are a lot of these automata — enough possibilities to hide a great many surprises.

Now let's consider the $Sz = 2$, $N = 9$ case: two-dimensional automata with Moore neighborhoods, for instance. In this case, there are $2^{2^9} =$

13,407,807,929,942,597,099,574,024,998,205,846,127,479,
365,820,592,393,377,723,561,443,721,764,030,073,546,
976,801,874,298,166,903,427,690,031,858,186,486,050,
853,753,882,811,946,569,946,433,649,006,084,096

automata rules. How can we understand such an immense number? Well, suppose we have a computer capable of examining a trillion of these automata each second. No, that's too slow. Suppose every elementary particle in the universe is a computer, examining a trillion automata each second, and suppose they have been doing this task since the birth of the universe. What fraction of the Moore automata will have been examined by now? Only about

$$\frac{1}{1000000000000000000000000000000000000000000000000}$$

Binary automata (S = 2) are complicated enough for many lifetimes of exploration, but allowing more states per cell increases the possibilities more rapidly still. Here is a single example: Consider one-dimensional automata with three states per cell (S = 3) and with the neighborhood of a cell consisting of the cell itself and one cell on either side (N = 3). For S = 3, we get $3^{3^3} = 7,625,597,484,987$ different automata rules, compared with only 256 for the S = 2 case. The number 2^{29} has 155 digits. By comparison, 3^{3^9} (the number of Moore automata with three states per cell) has 9392 digits. There are no sensible comparisons we can evoke to convey how many 3-state Moore automata exist.

Okay, this should have convinced you of two things: (1) even simple dynamical systems like Moore automata can have such a staggeringly large array of possibilities that all kinds of very strange behavior can remain forever unknown, and (2) to work in any reasonable way, we are going to have to restrict our attention to certain families of automata. Figure 9.5 gives a hint of the variations possible from just one-dimensional, binary automata with 5-cell neighborhoods. (S = 2, N = 5). See [115] for these and some other interesting patterns. The figure depicts time-lapse photographs of populations of on-cells (time increases in the downward direction), each starting from a random distribution, for three of the $2^{25} = 4,294,967,296$ possible rule combinations.

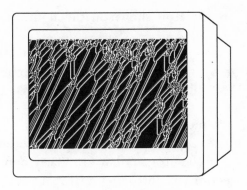

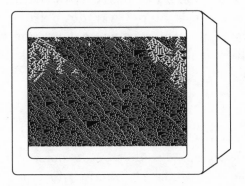

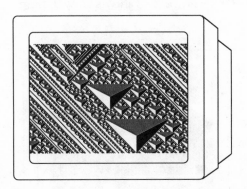

Figure 9.5 *Succeeding generations of on-cells from random initial distributions of on-cells for three different rules.*

The "netlike" automaton shown in the figure has the complicated rule

where each block means "If this configuration, then the central
cell is on in the next generation," and "otherwise off " is understood. The other two automata have similarly nonintuitive rules.

In the next section, we introduce a special type of symmetrical
automaton rule that is easy to codify and to which most of our
investigations will be limited. Despite their reduced flexibility,
these rules contain lots of unexpected behavior. For example,
look about two-thirds of the way down from the top in the
middle of Figure 9.6 The starting configuration in this figure is
three on-cells side by side. The "laughing skull" is somehow
built into this initial state and the governing rule.

Outer Totalistic Rules

Though it is less graphic to do so, it is sometimes convenient (as
we saw at the end of the last section) to represent on and off
states by 1's and 0's. Thus the rule   ■□□ *or* □□■ , which we
discussed earlier, might also be represented in an equivalent form
by (100) *or* (001). Figure 9.7 shows how a single on-cell evolves

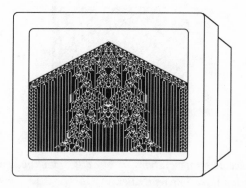

Figure 9.6   *A portrait of one of the authors painted in the generations of a
one-dimensional automaton.*

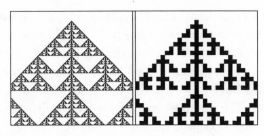

Figure 9.7   *Evolution of a single on-cell under the rule (010) or (001) or (100) or (111).*

under the slightly more complicated rule (010) *or* (001) *or* (100) *or* (111). (The righthand side of the figure is a magnification, allowing us to see individual cells turning on and off.) Note that *this* pattern of triangles within triangles is *not* that of the Sierpinski Gasket; here we see something rather more complex and interesting.

The rule (010) *or* (001) *or* (100) *or* (111) can be rewritten as follows: "If, in a 3-cell neighborhood, the central cell is on and has no on-neighbors, or is off and has one on-neighbor, or is on and has two on-neighbors, then it is on in the next generation; otherwise, it is off." Such a rule is called an outer totalistic rule because it can be stated in terms of the *total* of on-neighbors outside of the central cell, not in terms of *which* of the neighbors is on. (Of course, the size of the neighborhood and whether the central cell was on or off also count.) Thus the rule (101) is an outer totalistic rule, whereas (011) is not, because in the latter, which neighbor of the central cell is on is important.

We can imagine keeping track of outer totalistic rules with a little scorecard. For the one-dimensional, 3-cell neighborhood case, the (as yet blank) scorecard would look like Table 9.1a. The numeral above each pair of boxes, ⊞ , represents the total number of on-cells in the respective neighborhood outside of the central cell. The left box of the pair is reserved for the case *central cell on*, and the right for *central cell off*. This is how we record a rule. Suppose, for example, the rule is "If the neighborhood of a central cell that is currently on contains exactly one on-cell, the central cell will be on in the next generation" (the "otherwise off" part being understood). The scorecard corresponding to that rule is Table 9.1b. The "x" signifies that the central cell for the "any one on-cell" neighborhood configuration must be on for the central cell to be on in the next generation. If the scorecard had been Table 9.1c instead, then the rule would have read "If the neighborhood of a central cell that is currently *off* contains exactly one on-cell, that central cell will be on in the next generation." Similarly, the scorecard given as Table 9.1d corresponds to the rule "If a neighborhood of *any* central cell contains exactly one on-cell, that central cell will be on in the next generation." Only "x's" code for "turn on's"; no "x," no turn on in the next generation.

For one-dimensional, 5-cell neighborhood outer totalistic rules, a blank scoreboard looks like Table 9.2a. Scorecards for a two-dimensional von Neumann automaton with outer totalistic rules look the same. For a two-dimensional Moore automaton with outer totalistic rules, the blank scorecard looks like Table 9.3a.

(Figure 9.2 was generated by the Moore automaton with the outer totalistic rule given as Table 9.3b, whereas Figure 9.6 was generated by the one-dimensional 5-cell automaton with outer totalistic rule given as Table 9.2b).

Here's a word of caution: In the previous section we used the notation ⊞⊞⊞⊞⊞■⊞⊞⊞⊞⊞ to represent actual on- and off-cells. The present notation is meant to convey information about the total number of on-cells in the neighborhood of a given cell; it doesn't specify details of which cells are actually on. Don't confuse the two.

Returning to the rule (010) *or* (001) *or* (100) *or* (111)— equivalently, Table 9.1e—we find that when it is applied to a random distribution of on-cells, it produces the richly textured Figure 9.8. Figure 9.9 compares the results of applying the rule given in Table 9.1e to a single on-cell configuration (the topmost tiny triangle) and to a short *band* of randomly selected on-cells (the bottom portion of the figure). Though the colony emerges from such an initial state with microscopic irregularity, on a macroscopic scale (that is, large compared to the size of the starting band), it shows the same patterning as the colony that emerges from a single initial cell. In a sense, the coarse morphology of the colony emerging from a random start "remembers"

Table 9.1

	2	1	0
a	☐	☐	☐
b	☐	☒	☐
c	☐	☒	☐
d	☐	☒	☐
e	☒	☐	☒
f	☒	☒	☐

Table 9.2

	4	3	2	1	0
a	☐	☐	☐	☐	☐
b	☒	☒	☒	☒	☐
c	☒	☒	☐	☒	☒
d	☒	☒	☒	☒	☐
e	☒	☒	☒	☒	☒

Table 9.3

	8	7	6	5	4	3	2	1	0
a	☐	☐	☐	☐	☐	☐	☐	☐	☐
b	☐	☐	☐	☐	☐	☒	☒	☒	☐
c	☒	☐	☐	☒	☒	☒	☐	☐	☐
d	☐	☐	☐	☐	☐	☒	☒	☐	☐

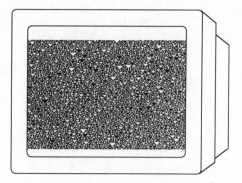

Figure 9.8   *Generations succeeding an initially random distribution of on-cells under the rule given as Table 9.1e.*

the microscopic rules governing its development. This is an example of *self-organizing behavior.*

Lest you think that rich texture always results from random initial distributions, consider Figure 9.10, which shows how an initially random distribution evolves under the rule given as Table 9.1f. After a brief initial flirtation with triangular pattern making, the configuration evolves into a set of bold contiguous bands that remains static in time. Under this rule, isolated cells tend to turn off, whereas clusters tend to stabilize. In short

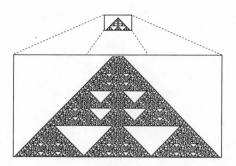

Figure 9.9   *Generations succeeding a single on-cell (top) and an irregular strip of on-cells (bottom) under the rule given as Table 9.1e. The generations in the lower picture show both the irregularity seen in Figure 9.8 on the small scale and the strict patterning seen in the upper picture on the large scale.*

Figure 9.10   *Not all random distributions evolve irregularly. Here, a random initial distribution becomes highly regular under the rule given as Table 9.1f.*

order, the initial state becomes balkanized, with segregated clusters favored — a kind of frozen state emerges. Under the previous rule we looked at, however, there is a balance between favoring clusters and favoring isolated individuals; the resulting spatial patterns are much more dynamic and complex in that case. (Is there a lesson beyond the mathematics to be learned from these simple examples?)

As might be expected, the 5-cell arrangement rules for one-dimensional automata are able to produce even richer behavior than the rules for the 3-cell arrangements that we have been examining. Figure 9.11 shows the evolution from a single on-cell

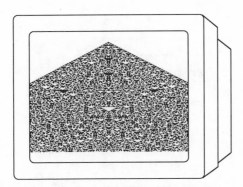

Figure 9.11   *Succeeding generations of a single on-cell under the outer totalistic rule given as Table 9.2c.*

configuration generated by the outer totalistic rule given as Table 9.2c. Instead of the lacy Gasket-like patterns we saw in Figures 9.4 and 9.7, here the growing pattern is much more textual — and, though symmetrical, also somewhat vaguely chaotic.

The outer totalistic rule given as Table 9.2d contains a surprise: when it is applied to an initially random configuration, dendritic structures can sometimes appear that are very reminiscent of diffusion-limited aggregation. Figure 9.12 shows some examples. Note that in the midst of one surprise there is yet another. Not only can this rule grow irregular, flowery protuberances, but in the dendrite on the right, a highly symmetrical zipper pattern suddenly emerges.

Let's briefly turn our attention to two-dimensional automata. The added dimension greatly enhances the richness of the possible evolutionary rules. Most rules are so complex for two-dimensional automata that reckoning how a starting configuration evolves by hand is distressingly tedious. Our best tool for studying such beasts is computer play.

For simplicity, we examine only von Neumann or Moore neighborhoods of a given cell (Figure 9.1). Recall that the scorecards for the von Neumann outer totalistic rules take the form of Table 9.2a and that those for Moore outer totalistic rules take the form of Table 9.3a.

As an example of the increased richness inherent in two-dimensional automata, Figure 9.13 shows every fourth genera-

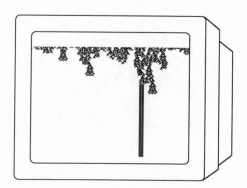

Figure 9.12   *Succeeding generations starting from a random distribution of on-cells under the outer totalistic rule given as Table 9.2d.*

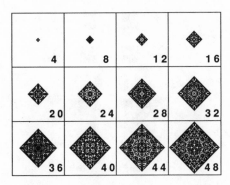

Figure 9.13   *A few generations of the colony of on-cells spawned by a single on-cell under the rule given as Table 9.2e.*

tion (stopping arbitrarily at generation 48) of the pattern evolving from a single on-cell for the rule given as Table 9.2e. Bear in mind that pictures of two-dimensional automata are snapshots of the cell space at a given instant, whereas the pictures we have displayed for one-dimensional automata are space (horizontal direction) versus time (vertical direction) records. Figure 9.14 gives the 48th generations (all starting from a single on-cell) of four different, though very similar rules. Note that the overall diamond shape is preserved from rule to rule but that the details of the on-cell patterns are markedly different.

Compare the results of Figure 9.14 with those given in Figure 9.15. Figure 9.15 shows four 48th-generation configurations

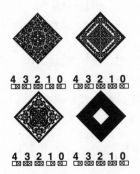

Figure 9.14   *The 48th generations of four automata with seemingly "nearby" rules. Each colony starts from a single on-cell.*

Figure 9.15   *The 48th generations of colonies under rules with the same total neighbors as in Figure 9.14, except now for Moore neighborhoods instead of von Neumann neighborhoods.*

(again starting from a single on-cell), employing rules with the same totals as in Figure 9.14 but now applying them to Moore neighborhoods instead of von Neumann neighborhoods. Note that the overall shape of these configurations is square; the overall shapes in both figures follow the structures of the respective neighborhoods (diamondlike for von Neumann, square for Moore). Note also that interesting configurations grow from the rules shown only because each contains the subrule "If the central cell has one on-neighbor, it is on in the next generation." (Can you see why that's the case?) Without such a subrule, the single on-cell configuration would never grow.

Only special rules permit interesting growth from a single on-cell. Similarly, special rules allow the development of interesting ordered structure from a random initial configuration of on-cells. See Figure 9.16 for an example. Note the very high degree of self-organization: in place of the random initial distribution of dots, large areas of horizontal and vertical line segments now appear. At least equally striking are Figures 9.17 and 9.18.

The Game of Life

The best-known cellular automaton is undoubtedly John Conway's *Game of Life*, which was first brought to the attention of a

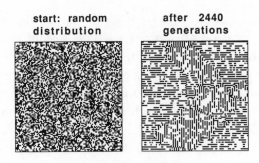

start: random
distribution

after 2440
generations

rule: 8 7 6 5 4 3 2 1 0

Figure 9.16   *The 2440th generation starting from a random initial distribution of cells for the rule given as Table 9.3c.*

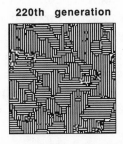

220th generation

rule: 8 7 6 5 4 3 2 1 0

Figure 9.17   *The pattern that evolved from a random configuration under the rule shown after 220 generations.*

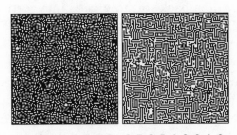

8 7 6 5 4 3 2 1 0   8 7 6 5 4 3 2 1 0

Figure 9.18   *The patterns that evolved from random configurations under the rules shown after 38 generations (left) and 58 generations (right), respectively.*

general audience by Martin Gardner [116]. In the first few years following publication of Gardner's article, a significant amount of computer time was spent playing Life. Life is not like computer chess, where a person engages a machine in a contest of strategy, nor is it a predecessor of video arcade games, where a person's reflexes compete with a machine's speed (appropriately hobbled). This game is played by the machine alone and has been described [117] as a "video kaleidoscope." The person specifies an initial configuration of live cells (we revert to the "living and dead" metaphor here) in a square array (the configuration space is two-dimensional with Moore neighborhoods), and the machine then follows a simple (binary, outer totalistic) automaton rule to produce generation after generation of living cells. Here is the rule: "If a cell is alive and exactly two or three of its Moore neighbors are alive, the cell remains alive in the next generation; if the cell is dead and exactly three of its Moore neighbors are alive, the cell becomes alive in the next generation. All other cells are dead in the next generation." In the notation introduced previously, Life corresponds to the rule given as Table 9.3d.

Some properties of Life are easy to check by hand. For example, an isolated live cell will die in the next generation (Figure 9.19a). So will a pair of adjoining live cells (Figures 9.19b and 9.19c). Four isolated live cells arranged in a square will persist forever (Figure 9.19d). Taken together, these four live cells are a fixed point of the dynamics. This pattern is called a *block*. Blocks are not the only stationary configurations. For example, Figures 9.19.e and 9.19f are also stationary. (Do you see why?) Three live cells arranged in a vertical line will change to three live cells in a horizontal line, then back to three cells in a vertical line, and so on (Figure 9.19g). This is a *blinker*. It is a 2-cycle of the dynamics.

Of course, still more complicated things can happen. For example, Figure 9.20 shows five generations in the evolution of a 5-cell configuration called a *glider*. After four generations this glider moves one square right and one square down (note the small, fixed circle placed in the upper lefthand corner of each frame for reference), recovering its original shape and orientation. If uninterrupted, it will continue to glide across the "Life field" forever.

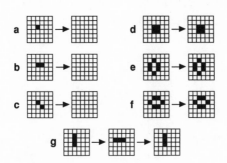

Figure 9.19   *Some elementary Life units.*

The top of Figure 9.21 is a 5-cell configuration called an *r
pentomino*. (The "r" comes from the shape, the "pent" from the 5
cells.) (The scale of the figure is much larger than the pictures
displayed in Figures 9.19 and 9.20, and the cross hatching has
been removed.) Unlike the 5-cell glider of Figure 9.20, the r
pentomino challenges an analysis by hand. It evolves into an
extraordinarily complicated colony, an example of which is seen
in the lower part of Figure 9.21; the pattern shown is the colony
that evolves from the r pentomino after 150 generations. Observe that at this stage, three gliders have been produced (they
are set off in circles), one moving northeast and two moving
southeast. Will the number of live cells in this colony continue to
grow forever (in an infinite field)? In Gardner's original column
describing Life, Conway offered a $50 prize to anyone who
either could produce a configuration that would grow without
bound (a glider skimming across the Life field is not growing; it's
just moving) or could prove there is no configuration that grows

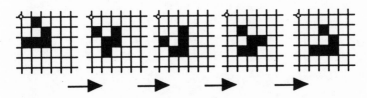

Figure 9.20   *A glider moving diagonally in the southeast direction.*

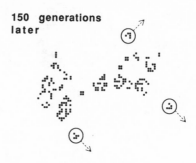

Figure 9.21 *The evolution of an r pentomino.*

forever. The r pentomino does not answer Conway's question. After 1103 generations (on an infinite field), it settles down to a stable picture that includes six gliders departing into the distance. At the very least, the r pentomino is a cautionary lesson that much patience may be required to determine the long-time behavior of a configuration.

The $50 prize eventually was won by William Gosper and five other MIT students for a configuration called a *glider gun* (see Figure 9.22). Every 30 generations the glider gun spits out a *new* glider, and all the gliders that are produced move off in a steady stream to the southwest. The figures between the two blocks are called *shuttles*; their fate is to bounce back and forth between the blocks. When they collide, the shuttles interact to produce a glider. Clearly, on an infinite Life field this pattern will grow without limit. In addition, Gosper and his co-workers showed that 13 gliders can be arranged so that they collide and form a

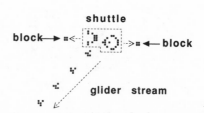

Figure 9.22 *A glider gun spitting out a stream of gliders.*

glider gun. Starting from this idea, Conway was able to show that there are *self-reproducing* Life patterns [118].

Conway has made a bold conjecture. Suppose we have an infinite Life field, with an initially random distribution of live cells. Almost any configuration, no matter how complex, will occur somewhere in this initial field. Thus somewhere in the infinite initial distribution, copies of Conway's self-reproducing configuration are likely to occur. Furthermore, variations in Conway's construction can move across the Life field, encountering other elements that are growing from the initial distribution. The moving configuration can be modified to send out properly aimed pairs of gliders that will collide some distance away and send a glider back. If this glider does not return, then the pair of gliders that went out must have encountered another Life pattern. That is, somewhere in the infinite Life field, there will be self-reproducing, moving configurations that can "see." Some of these will encounter other elements of the field, causing modifications that can be thought of as mutations. Most mutations will undoubtedly cause "death"; but some will be favorable, and a better-adapted "creature" will result. In this way, the inhabitants of the Life field can experience a sort of evolution. Given enough ticks of the Life clock, intelligent, self-aware creatures could evolve and move purposefully across the field. Of course, it's not clear how one would store a truly infinite Life field or update its configurations, but you see the implications. (Pushing Conway's conjecture to its logical limit, one wonders if *we* might be critters in some cosmic simulation of a variant of the Game of Life?)

How Can Automata Behave?

We have suggested that the video feedback demonstration we discussed in Chapter One can be thought of as a kind of cellular automaton. We noted in Chapter One that video feedback can produce qualitatively distinguishable effects, which we summarize here.

1. The screen flashes, but the pattern that is produced rapidly extinguishes, and thereafter the screen remains dark.

2. A spontaneous flash occurs and rapidly stabilizes into an unchanging or, perhaps, periodically pulsating blob.

3. The screen flashes on, and light dances across the screen in some erratic fashion and disappears. A bit later, another flash leads to a different dance; this sequence of events repeats only after very long intervals (if at all).

4. The spontaneous flash evolves into "organic" structures, moving patterns with extraordinarily complex spatial structure.

We also observed that control settings for effects 1, 2, and 3 are relatively easy to find by trial and error but that settings for type 4 are more elusive and seem to be buried between those for types 2 and 3. The characteristics of the phenomena found in video feedback are consistent with how cellular automata in general are believed to behave.

In the early 1980s, Stephen Wolfram [119] observed that from random starting configurations, automata eventually produce one of four types of behavior:

Class I	Everything dies (or everything lives).
Class II	The pattern of live cells repeats exactly after some number of generations.
Class III	The pattern of live cells grows in a chaotic fashion.
Class IV	The pattern of live cells grows and contracts in a complicated way, exhibiting both fixed and "moving" patterns.

Class IV is the most interesting, and the most rare. Conway's game of Life is a class IV automaton, gliders being examples of the moving patterns. Is there some way we can understand how this variety of patterns arises?

In the mid-1980s, Christopher Langton [120] discovered that Wolfram's classification scheme could be made somewhat quantitative by computing a simple ratio. This parameter, now widely referred to as Langton's lambda, λ, roughly determines automata behavior in much the same way in which the control parameter, s, determines the dynamical behavior of the logistic map. For binary automata, λ is defined by

$$\lambda = \frac{\text{number of neighborhood configurations producing live cells}}{\text{total number of neighborhood configurations}}$$

Because every automaton with $\lambda > 1/2$ corresponds exactly to an automaton with $\lambda < 1/2$ after the roles of "live" and "dead" cells have been interchanged, we consider only values of λ that lie between 0 and $1/2$. Langton found that as λ increases from 0 to $1/2$, automata generally go through class I, class II, class IV, and finally class III. That is, class IV automata are on the boundary between the "predictable" classes I and II, and the chaotic class III. This transition, where class IV automata exist, occurs at a "critical" value λ_c. Langton also observed that as λ approaches λ_c, automata exhibit longer and longer "transients"—that is, they take more and more time to settle down to their eventual behavior.

Perhaps now you can see the crude analogy between λ and the logistic map's s. At low s values, the associated attractors are fixed points (class I). At higher values, periodic attractors emerge (class II). At higher values still, the dynamics goes into chaos (class III). At the critical value of s, just before the emergence of chaotic behavior, the period of the cyclic attractor can be very, very long. For these cases, arbitrary initial states often take quite a while to settle down into the ultimate periodicity. (Though the one-dimensional logistic map cannot show complex, moving spatial structures, the long transient lead into periodicity is reminiscent of the behavior of class IV automata.) There is also a vague analogy between λ and temperature. Classes I and II seem frozen like crystals, for example (at low temperature). Class III, on the other hand, is like the turbulence of a swirling liquid (at high temperature). Class IV behavior suggests the transition from solid to liquid or vice versa (at the critical freezing temperature). In either analogy, class IV is behavior far from equilibrium.

In general, Langton's λ more aptly describes the behavior of automata when the number of neighborhood configurations is large. It works reasonably well, however, even for 5-cluster binary automata. For example, the automaton with the rule

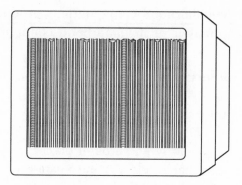

Figure 9.23   *A frozen, class II, one-dimensional automaton.*

(that is, "If ■■□■ , or ■■□■ , or . . . , or □□□■ , the central
cell is alive; otherwise, it is dead") exhibits class II behavior (see
Figure 9.23) and has $\lambda = 10/32$. Adding ■□□□■ gives a
class IV automaton (Figure 9.24) with $\lambda = 12/32$. Adding ■■□■
□■□■ to this gives a class III automaton (Figure 9.25) with
$\lambda = 14/32$.

We should mention that the situation is a bit more complicated than this example suggests. Starting with different rules,
the transition from class II to class III can occur at different
values. Also, for some rules it seems possible to go from class II
directly to class III without passing through class IV. The situation is depicted schematically in Figure 9.26. This figure shows in
a cartoonish manner the collection of all possible rules for an

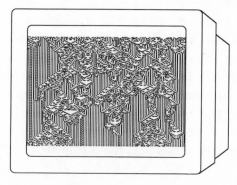

Figure 9.24   *An organic, class IV, one-dimensional automaton.*

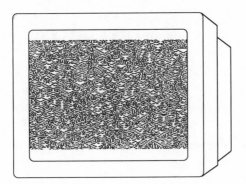

Figure 9.25   *A chaotic, class III, one-dimensional automaton.*

automaton of given dimension and neighborhood. The automaton's behavior tends to partition into regions of rule space that are parameterized by Langton's lambda, as shown. Although many exceptions can be found, the Langton prescription seems to be a useful general guideline.

Class IV automata have some remarkable properties. For example, any computer program can be simulated by a class IV automaton. There are clever arguments establishing this for the game of Life by showing how proper arrangements of glider guns can execute the basic logical operations "and," "or," and "not" of computer programming. The fixed patterns and moving pat-

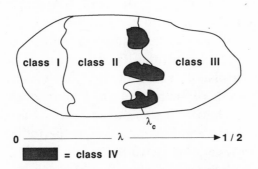

Figure 9.26   *Schematic diagram of a typical automaton rule space. Every point within the major blob represents a possible rule. Associated with each rule is a Langton parameter, λ. As λ increases, the long-term behavior of the automaton changes from class I to class III. Class IV behavior is depicted as islands nested near λ_c on the boundary between class II and class III.*

terns in class IV automata provide the ability to store and to transmit information, the two main tasks a computer must perform. Langton has speculated that living systems are class IV automata, balanced precariously between static behavior (death) and chaotic behavior (dissolution). He suggests that living creatures maintain themselves in class IV by exploiting the very long transients near λ_c. Within this conjecture, life is seen as an "impermanent phenomenon" teetering on the edge of chaos.

Life, Sand Piles, and 1/f Noise

When the Game of Life is played on a finite, wraparound field, initial configurations that burst into long periods of growth eventually settle into periodic states, after which nothing much happens. These periodic states are usually populated with stationary and blinking microclusters such as those we discussed when we described the Game of Life. But the quiescent end-states of Life are very dissimilar to those of periodic, class II automata. Periodic automata are frequently stable. If perturbed by the addition of a cell at some random location in the field, these automata undergo, at most, slight configurational alterations that quickly evolve back to essentially the same periodic state.

Life is different. Suppose we have a quiescent Life state consisting of a dilute scattering of small stationary and periodic clusters. Suppose we perturb this state by inserting a live cell somewhere at random. Most of the time the new cell will simply die, because most of the time its site will have no surrounding neighbors (and isolated cells die under the rules of Life). Sometimes the new cell will be adjacent to a microcluster. Often that perturbation will initiate a short-lived flurry of activity that dies out, or evolves into another stationary or periodic microcluster. Sometimes, however, the perturbation can explode into an extended period of growth, the results of which can sweep completely across the Life field. An example is shown in Figure 9.27.

Quiescent states of Life are apparently far from stable equilibrium. In physical terms, we would say that embedded in the Life rules — because both birth and death are possible — is an implied

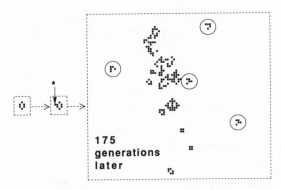

Figure 9.27   *A single-cell perturbation (∗) to a stationary cluster produces an explosion of growth. After 175 generations, the colony's size is many times that of the starting cluster, and gliders (circled) are being spawned that have the potential of spreading the evidence of the perturbation over a huge area.*

interaction of a colony with its surroundings. In effect, a colony is immersed in an inexhaustible reservoir of such resources. For some cluster configurations, this swapping is inefficient. Single cells, for example, are incapable of harnessing the potential of this interaction. But for other configurations — clusters of cells acting *collectively* in the right way — the interaction with the external world leads to sustained activity. The quiescent yet strangely fecund states of Life do not require "fine tuning." It is not the case that only a few carefully specified initial configurations will lead to such end-states. Almost any initial state will do. In that sense, the typical quiescent states of Life are said to be *self-organizing*. Their inevitable appearance is encoded in the rules of the Game.

Small perturbations of quiescent Life states can touch off fluctuations of wildly different sizes and durations. A similar situation occurs when a vapor is carefully cooled to exactly the right temperature and pressure. At one very special point in the cooling, the vapor — just on the verge of condensing into a liquid — becomes opalescent (not clear, not opaque — sort of milky). This is because at that point, the container is filled with isolated vapor molecules as well as denser, almost liquidlike clusters of molecules of all sizes. (Different-sized clusters scatter light differently. With all those different sizes present, nearly all light that is shone on the container is scattered; it's hard to get any to pass cleanly through.) These fluctuating clusters constantly form, break apart, reform, ... The stuff in the container is schizophrenic: should it be vapor or should it be liquid? Such a thermodynamic state, characterized by fluctuations on all length scales, is termed a *critical state*.

Thermodynamic critical states are difficult to achieve. They require fussy control of parameters. They are not self-organizing. To differentiate the robust critical behavior (that is, fluctuations on all length and time scales) of automata such as Life from the fragile critical behavior of thermodynamic systems, Per Bak [121] and others have developed the notion of *self-organized criticality* (SOC). Bak and his various colleagues suggest that self-organized criticality is ubiquitous, appearing, for example, in fluid turbulence, the burning of forest fires, the onset of earthquakes, and the trading of commodities. An especially transparent plausible candidate for SOC behavior is an ordinary sand pile.

Imagine dropping dry sand a grain at a time onto a small dish, as shown in Figure 9.28. At first, the sand spreads out into a broad, flat smudge. As the process continues, the grains pile up into a conical mound. As the cone steepens, avalanches begin to slide down its slopes, eventually spilling over the edge of the dish. These avalanches serve to maintain the slope of the mound at a more-or-less constant angle, the so-called *angle of repose*. The angle of repose is determined by the physical properties of the sand (grain size and roughness), not by the size of the mound (a sand pile that fills the entire table has about the same angle of repose as the pile on the small dish). How the pile was formed is

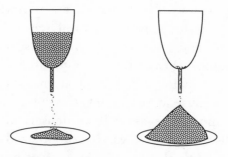

Figure 9.28   *Building up a sand pile.*

also irrelevant. (You could make a tall cylinder of wet sand, let it dry, and after the cylinder slumped down, find the same angle of repose that emerges when grains are dribbled from above.) Somehow, without regard to scale or to the details of how it was made, a sand pile organizes itself into a critical shape. Adding more sand to the pile causes rearrangements, but after all the activity settles down, the critical shape is reestablished (more or less). It is useful to think of sand pile building as a dynamical system in which added grains and induced avalanches conspire to produce a roughly invariant pile shape. In this view, *the critical state is an attractor for the dynamics.*

Events within self-organized critical (SOC) systems can occur over wide ranges of lengths and times. In these systems, complex spatial structures give rise to complex time behavior. In fact, it has been suggested that this characteristic of SOC systems may be the long-sought explanation for where 1/f noise originates. To get a glimpse of how this might happen, let's consider a variant of a model "sand pile" first investigated by Bak, Tang, and Wiesenfeld [122]. A physical realization of the model is similar to the "ant box" contraption depicted in Figure 9.29. At the left end of the box there is a impenetrable barrier, and at the right end there is a hole through which grains can fall. Sand is confined between two flat, vertical plates. The left side of the pile is bounded by a vertical wall, whereas the right side is free to fall through a hole in the bottom of the container. We imagine that the top of the

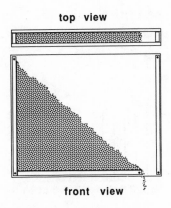

Figure 9.29   *A sand pile in an ant box.*

box is open and that we can dribble grains into the box from above.

Let's cast the model in the language of cellular automata. The space of the automaton is one-dimensional (that is, the opening between the front and back plates in Figure 9.29 is very narrow) and consists of some number of cell sites, L. Each site in the space can have any one of a number of states, given by some positive number — not necessarily just 0 or 1. (The automaton is *not* binary.) This state represents the number of grains of sand stacked in a column above the cell site.

A slightly abstract version of Figure 9.29, more in keeping with the cellular automaton model, is shown in Figure 9.30. The active sites in the space are numbered from 1 to L (L = 10 in Figure 9.30). Sites 0 and L + 1 are boundaries. The height of the column at site 0 is always adjusted to equal that at site 1. This end of the pile corresponds to the impenetrable barrier (or, equivalently, to the center of a symmetrical pile the lefthand half of which you can't see). The height at the other end of the pile, at site L + 1 (remember, L + 1 is 11 in Figure 9.30), is held fixed at 0. This end corresponds to a hole (or edge).

The automaton has dynamical rules. Let $S(i)$ be the height of the column at site i minus the height of the column at site i + 1. The quantity S is a measure of the steepness of the pile at each site. (By definition, the steepness at the left end is 0; $S(0) = 0$; and at the hole end, $S(L) =$ the height of the column at L.) The automaton rule is

If $S(i)$ exceeds some critical value, S_c (S_c is positive), then N_t grains are transferred from site i to site i + 1 in the next generation.

0 1 2 3 4 5 6 7 8 9 10 11

Figure 9.30   *The ant box with sand columns at discrete sites.*

For example, say S_c is 3, N_t is 2, and S(5), in Figure 9.30, is 4. Then, in the next generation, 2 grains are removed from site 5 and transferred to site 6. Note that any grain moved from site 1 to site 2 causes 1 grain to be removed from site 0 and that any grain moved from site L to site L + 1 is removed from the sand pile. Note also that requiring S(i) to be greater than S_c for grain transfer restricts grains to move from left to right. The rule is applied over and over until no more moves can be made. At that point, no local steepness is greater than S_c.

Suppose we start with some quiescent configuration; (that is, all S(i) are less than or, at most, equal to S_c). Imagine perturbing this quiescent state by dropping a single grain on one randomly selected site in the box. What is the effect? Let's take a specific example. Suppose $S_c = 3$. Suppose site 5 has 13 grains, site 6 has 10 grains, site 7 has 7 grains, and site 8 has 5 grains. In this case, the steepness at site 5 is $13 - 10 = 3$, that at site 6 it is $10 - 7 = 2$, and at site 7 it is $7 - 5 = 2$. One grain is added at site 6. Now, the steepness at site 5 is $13 - 11 = 2$, that at site 6 is $11 - 7 = 4$, and at site 7 it is still $7 - 5 = 2$. The steepness at site 6 is now too large, so grain rearrangements occur. If $N_t = 2$, for example, then in the next generation, 2 grains shift from site 6 to site 7. Although such a shift stabilizes site 6, it may destabilize other surrounding sites. Accordingly, if no grains are shifted in the next generation into site 5 and no grains are shifted out of site 8, then the steepnesses at sites 5, 6, and 7 will be 4, 0, and 4, respectively. (Work through the arithmetic.) Thus stabilization of site 6 destabilizes sites 5 and 7. This process of interconnected stabilization and destabilization can cause avalanches sweeping, in time, over the entire sand pile. (*Note*: As in all cellular automata calculations, you need two copies of the space: the present generation and the next. You update the entire cell space by using the dynamical rules. For the sand pile example, the height of a column at a given site in the next generation is adjusted by what comes into the site versus what goes out when the rule is applied in the present generation to that site and its neighbors.)

When N_t is exactly 1, the quiescent state that eventually forms after many perturbations is one in which the steepness is exactly S_c at every site. (It's an interesting exercise to prove to yourself that this is true.) When N_t is greater than 1, though, surprisingly complicated avalanches can be produced. Most avalanches are very small, but some can grow quite large. Even for as small a sand pile as $L = 30$, avalanches involving the motion of about 1000 grains can be observed. Such avalanches have similarly long durations. Now suppose the sand pile is very large (so that avalanches don't collide with each other) and grains are dropped on it at random moments and at randomly selected places. What is the nature of the total number of grains moving in the pile at any instant? Here's this big sand pile with avalanches constantly rumbling over it. Most of the time, the individual avalanches are small and so is their sum. Rarely, a number of big avalanches occur at different places simultaneously. The sum is then big. Thus the total activity in the sand pile fluctuates from instant to instant; it is a noisy signal. What kind of noise is it?

The relaxation of the sand pile is driven by height differences. If the steepness is too great at one spot, then grains move in such a way as to reduce the steepness. In that sense, the model is similar to diffusion: if there are a lot of ink molecules in one spot in a glass of water, they tend to diffuse around, thus reducing the differences in concentration of the ink from place to place in the glass. Diffusion is Brownian motion, and Brownian fluctuations have a spectrum that drops off like $1/f^2$. The same is true of the avalanche activity fluctuations in the model described here. The sand pile (and presumably lots of similar systems) acts like a nonlinear amplifier. You put in the same, small loudness at all frequencies, but what comes out is a lot of loudness at low frequencies. Note that if the sand pile were not in a critical state, its output would faithfully follow the input. It would be equally weak at all frequencies. But in the critical state, small inputs can amplify into huge outputs — through infrequently.

Now, here's an interesting twist to the model. Suppose that the front plate in the ant box of Figure 9.29 is actually a coarse screen. And suppose that when grains topple from one site to the next, some fraction of the moving grains fall out of the screen and are lost. How does that change things? Well, suppose that each time grains topple, they all fall out. Then no sustained avalanches will be spawned, and the spectrum of the total activity in the pile will mimic the way in which it is being perturbed. If grains are dropped in uniformly randomly, then the activity in the pile will also be uniformly random. In this event, the activity in the pile has a white noise spectrum ($1/f^0$). Thus at one extreme, when the fraction of grains falling out of the screen is 0, the total avalanche activity spectrum is $1/f^2$, whereas at the other extreme, when the fraction falling out is 1, the spectrum is $1/f^0$. Clearly, by adjusting the fraction of grains that fall through the screen to be somewhere between 0 and 1, we can make the activity spectrum $1/f^1$. Consequently, $1/f$ noise may be a natural consequence of perturbing an SOC system with white noise, as long as dissipation (loss of some kind) accompanies the relaxation of the system. [We note again Mandelbrot's observation (first mentioned in Chapter Six) that the electrical noise at the periphery of our nervous system (at the stimulus end) is white but that at the system's center (at the response end) the electrical noise is correlated, and his associated speculation that this electrical activity underlies human esthetic choice. To push the speculation one step further, in light of what we have just observed about sandpiles, we wonder to what extent esthetics is simply a direct consequence of living matter having the structure of self-organized criticality.]

Though this is a somewhat controversial view, Bak argues that many natural fractals are generated by some form of self-organized criticality: in this picture, fractals and $1/f$ noise are the space and time signatures of self-organized criticality. The idea that the same mechanism could give rise to fractal structures in space and in time is quite appealing. And it shares some features of Langton's "evolution to the edge of chaos," though the precise relationship between the two is not clear.

Neural Networks

The final automaton we will consider is called an artificial *neural network* [123]. An artificial neural network is a highly simplified model of a small, primitive brain. Despite their relative simplicity, model neural networks have been developed that perform astonishing cognitive feats ranging from voice and pattern recognition to robotic control and simple decision making. Their potential for practical application has excited a surge of research interest in recent years. At this writing, there are hundreds of documented uses of neural networks in business and industrial situations, and it is likely that neural networks will find even wider utilization in the next few decades. Neural networks are based on the observation that cellular automata are capable of "computing." Here's an example. Consider a one-dimensional, 3-cell automaton with the rule "Regardless of the current state of the central cell, if neither of its two neighbors is on, then the central cell state is off in the next generation; otherwise it is on in the next generation." This automaton "computes the logical inclusive OR function." What does all *that* mean?

Let's start with the inclusive OR function. This function is equivalent to the phrase "p *or* q." In classical logic, the phrase p *or* q is said to be true when p is true, or when q is true, or when p and q are both true, but it is false when p and q are both false. This truth/falsity situation is often summarized in a *truth table*, where T represents true and F false.

Inclusive OR:

p	q	p or q
T	T	T
T	F	T
F	T	T
F	F	F

Now imagine the 3-cell automaton redrawn as shown in Figure 9.31. In the right-most piece of the figure, the central cell is pulled away from its neighbors to emphasize its different role; it's designated as the output cell. The two neighbors are called

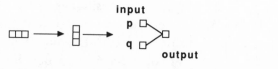

Figure 9.31   *A 3-cell automaton redrawn to appear as an input/output device.*

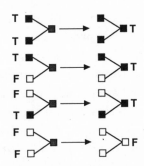

Figure 9.32   *The inclusive OR truth table is computed by the rule "the central cell is on in the next generation if at least one of its neighbors is on in the present generation."*

the input cells, and for specificity, the upper one is called the p input and the lower the q input. To proceed, let "true" be associated with "on," and "false" with "off." Figure 9.32 shows how the rule expressed in the first paragraph computes the truth table for inclusive OR. The column to the left in Figure 9.32 represents different input, or starting, configurations. The column to the right shows the corresponding output configurations that occur after a single application of the rules. The gray cells on the left indicate that it doesn't matter what state the output cell is in initially.

There is no difference between the outputs induced by (p = T, q = F) and by (p = F, q = T) in the logic functions inclusive OR, exclusive OR, and AND. In other words, the automata rules associated with each function are insensitive to which cell is called p and which q—they are outer totalistic rules. This is not the case for the logical function IMPLIES. The truth table for p *implies* q is

Implication:

p	*q*	*p* implies *q*
T	T	T
T	F	F
F	T	T
F	F	T

This function says that a true premise can have only a true conclusion but that a false premise can imply anything. Now $(p = T, q = F)$ and $(p = F, q = T)$ no longer give the same truth value. The associated automaton rule cannot be outer totalistic; it has to differentiate between the p input and the q input. One form might be "If the p cell is on and the q cell is off, the output cell is off in the next generation; otherwise, it is on."

In outer totalistic rules, the next state of a given cell depends on a sum of state values of the cells in the given cell's neighborhood. A kind of outer totalistic rule that permits a 3-cell automaton to compute IMPLIES uses *weighted connections* between input and output cells. In this picture, the connections between p and the output cell and between q and the output cell are different. To implement this idea, let the on state (= true) have an *activity value* 1 and let the off state (= false) have an activity value 0. Assign to the p-to-output connection a weight W_{po} and to the q-to-output connection another weight W_{qo}. The activity of the output cell — that is, the output state — will be determined by the input to the output cell from its neighbors. In this case, that input is taken to be the weighted sum

$$I_o = W_{po} \cdot A_p + W_{qo} \cdot A_q$$

In the latter expression, the A's are the activities of the respective input cells; multiplying the A's by weights serves to transfer the activity of the input cells to the output cell. We then need a rule that translates I_o into the desired output cell activity — that is, into a "true" or a "false." There are many ways to do this. One simple choice is $W_{po} = -1$ and $W_{qo} = +1$, with the automaton rule "If I_o is negative, the output cell is off in the next generation; otherwise, it is on." Let's see how that works:

p	q	I_o	A_o
T	T	$(-1) \cdot (1) + (1) \cdot (1) = 0$	TRUE
T	F	$(-1) \cdot (1) + (1) \cdot (0) = -1$	FALSE
F	T	$(-1) \cdot (0) + (1) \cdot (1) = +1$	TRUE
F	F	$(-1) \cdot (0) + (1) \cdot (0) = 0$	TRUE

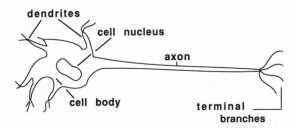

Figure 9.33   *A schematic sketch of a real neuron.*

This assigning different weights to the different connections and utilizing weighted sums leads to an automaton model that is vaguely similar to the way actual nerve cells in the brain—that is, real neurons—"compute." Figure 9.33 depicts a neuron. The brain and central nervous system are filled with these cells, and they are among the largest in the organism in which they are present. Neurons tend to form extraordinarily complex networks. In the human brain, for example, the dendrites of one neuron (of which there may be 10^{10} or so) may make close contact with the terminal branches of hundreds to millions of other neurons; its terminal branches, in turn, connect to many other neurons. Dendrites are constructed in such a way that they are able to receive chemical stimuli (for instance, molecules of the substance acetylcholine) emitted from special transmitter vesicles in terminal branches. Under some conditions, the accumulated effect of the many dendritic chemical stimuli changes the electrical properties of the neuron; an electrical pulse then results and sweeps down the axon. This pulse can trigger the release of transmitter from the excited neuron's terminal branches. The strengths of the connections between the terminal branches and successive dendrites need not be identical. Thus information, coded electrochemically, can be passed from neuron to neuron and processed in between. Such a network is a wonderfully subtle and flexible (though, by silicon standards, slow) computer. Among its many talents, it can recognize aural and visual patterns, calculate, manipulate symbols, strategize, philosophize, and experience feelings. Most amazingly, it can learn.

Although the 3-cell automata we have talked about are capable of some very primitive computing, a reasonable automaton model of any real neural network would involve many neurons interconnected via differentially weighted links. Furthermore, such an automaton has to be able to learn: it has to be able to discover its own rules. One general schematic for a model neural network is given in Figure 9.34. In this scheme the artificial neurons are grouped into three functional clusters: a layer of input neurons, a layer of output neurons, and perhaps several "hidden" layers of neurons whose purpose is to process information flowing from the input to the output. The numbers of neurons in these layers need not be the same, nor do all neurons in successive layers have to be connected. These considerations are aspects of the "wiring diagram" or the "hardware." Choices of the numbers of layers, the numbers of neurons in each layer, and the geometry of interconnection are said to make up the "architecture" of the network. In the language of cellular automata, these choices coincide with the definitions of the space of the automaton and the relevant neighborhoods.

What a network will compute depends in part on its architecture and in part on its "software" — that is, the specific choice of weights for each of its internal connecting links, and how a summed input to a neuron changes its level of activity. (The latter is equivalent to definition of the automaton's state and the assignment of the dynamics rules.) No doubt some computing ability in a real neural network comes courtesy of the molecular machinery that builds the network; some programming is "hardwired" by genetics. On the other hand, almost all organisms evince at least some rudimentary ability to learn. Capturing that feature of real neural networks is what allows model networks to execute desired tasks.

How can an automaton model of a neural network be made to learn? One answer that has sometimes been found to be effective is called supervised training, wherein the network is shown, and then quizzed on, a list of inputs and "correct" associated outputs. It goes as follows: Pick the architecture of a model network like that shown in Figure 9.34. Select a rule that translates inputs to a cell into cellular activity. Sprinkle random weights throughout the network. (A little variety to start with seems to be

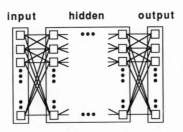

Figure 9.34   *A schematic representation of an artificial neural network.*

necessary; it has been found by trial and error that starting off with equal weights everywhere hangs up the learning process.) Select *a training set*—a collection of input/output pairs. Apply the first input to the input neurons. Allow an output to be computed.

A word about how computation occurs in the architecture of Figure 9.34 is warranted. As in our 3-cell examples, information flows from input to output. When there are multiple hidden layers, the computation is done sequentially from layer to layer. The states of the cells in the layer adjacent to the input layer are changed first, then those in the layer adjacent to that are changed, and so on, until finally those in the output layer are changed. With this architecture, that is the end of the computation.

Because the network is loaded initially with a random nonsense program (the random sprinkling of weights), the first output that is computed typically bears little relation to the desired output. Training proceeds by "taking the difference" between the desired output and the actual output and adjusting weights in response to that "error." (One algorithm for adjusting weights, called backpropagation, is described in the next section.) "Taking the difference" means just subtracting if the output is a single number; when the output is more complex, such as a set of numbers, some other appropriate rule has to be invoked. Once all the weights have been changed, a new program has been created. Let the network compute a new output using this new software, but with the same input as before. Generally, the second computed output is closer to the desired output but still not equal to it. If it is not closer, repeat the error feedback process and try again. Eventually the output error will become tolerably small (you have to specify ahead of time how much error you're willing to tolerate) if things work the way they should. Go on to the second pair of the training set. Starting with the weights that compute the first output properly, compute the output for the second input. Generally, the agreement between computed and desired output is poor, so the whole process invoked for the first input/output pair will have to be repeated for the second pair. And so it goes. Go completely through the whole training set once. Now, the weights arrived at to compute the final output of the training set probably won't do very well for the first input/output. (The network does okay on its last lesson, but by that time it has "forgotten" its first lesson—though not completely.) The entire regimen is repeated again and again. If there is convergence (and there are lots of reasons why there may not be), then after what may well be many iterations through the entire training set, weights will be arrived at that compute a tolerable output response to every input in the list. At this point, the network is said to be trained. It is able, within some tolerable level of error, to associate correct outputs with their respective inputs from the training set.

How Backpropagation Works

The basic idea of backpropagation is very simple. We illustrate it with a first step in building a neural network to perform the logical operation IMPLIES. Consider a network with two input neurons, p and q, both of which are connected to two neurons labeled r and s, in a hidden layer, and a single output neuron o that, in turn, receives input from both r and s. The wiring is shown in Figure 9.35. The hidden-layer and the output neurons both receive input and transmit output. We require a rule that converts input level to output level. For this example, we take the rule "If the summed input to a neuron is negative, the output of that neuron is 0 (=false); otherwise, it is 1 (=true)." Suppose the initial weights are

$$W_{pr} = 2.5 \qquad\qquad W_{ps} = -1.4$$

$$W_{qr} = 1.2 \qquad\qquad W_{qs} = 2.1$$
$$W_{ro} = -1.6 \qquad\qquad W_{so} = 1.4$$

The architecture with these weights is shown on the lefthand side of Figure 9.35. Now suppose we choose from the IMPLIES training set the input/output pair (p, q)/o = (F, T)/T; that is, p = F = 0, q = T = 1, and o = T = 1. First we compute the inputs to neurons r and s, using the activation values $A_p = 0$ and $A_q = 1$. The input to r is

$$W_{pr} \cdot A_p + W_{qr} \cdot A_q = (2.5) \cdot (0) + (1.2) \cdot (1) = 1.2, \quad \text{so } A_r = 1$$

whereas the input to s is

$$W_{ps} \cdot A_p + W_{qs} \cdot A_q = (-1.4) \cdot (0) + (2.1) \cdot (1) = 2.1, \quad \text{so } A_s = 1$$

We use these activation levels to determine the input to neuron o:

$$W_{ro} \cdot A_r + W_{so} \cdot A_s = (-1.6) \cdot (1) + (1.4) \cdot (1) = -0.2, \quad \text{so } A_o = 0$$

That is, with these weights our neural net thinks that "false implies true" is false. This is wrong, and the "error" at neuron o, e_o, is defined to be the difference between the desired output (1) and the actual output (0). That is,

$$e_o = 1 - = 1$$

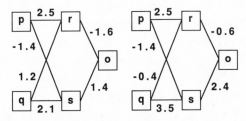

Figure 9.35 *A 5-neuron network with two input neurons, two hidden-layer neurons, and a single output neuron.*

Backpropagation takes its name from the next step in the training process; the error at neuron o is "propagated back into the network." The backpropagation algorithm assumes that the error at o propagates back to r and s through the respective weights connecting them to o. That is,

$$e_r = e_o \cdot W_{ro} = 1 \cdot (-1.6) = -1.6$$

and

$$e_s = e_o \cdot W_{so} = 1 \cdot (1.4) = 1.4$$

In this example, the error propagates no further back. In a network with additional hidden layers, the error would propagate layer by layer backward in a similar fashion until all hidden neurons were assigned an error value.

Now, the errors that are assigned in this manner are used to reassign weight values. Changes to the weights, W_{ro} and W_{so}, that are closest to the output neuron (designated ΔW with appropriate subscripts) are defined as the product of the error e_o and the outputs of neurons r and s, respectively.

$$\Delta W_{ro} = e_o \cdot A_r = 1 \cdot 1 = 1 \quad \text{and} \quad \Delta W_{so} = e_o \cdot A_s = 1 \cdot 1 = 1$$

The new weights are the sum of the respective old weights and these changes.

$$W_{ro} = -1.6 + 1 = -0.6 \quad \text{and} \quad W_{so} = 1.4 + 1 = 2.4$$

(To keep the notational complexity to a minimum, we're using the same symbols, but we mean these to be the changed weights.)

Similarly, the changes in the weights W_{pr}, W_{qr}, W_{ps}, and W_{qs} are defined as the product of the errors e_r and e_s with the appropriate outputs of the neurons $_p$ and $_q$. Thus,

$$\begin{aligned}
\Delta W_{pr} &= e_r \cdot A_p = (-1.6) \cdot (0) = 0 \\
\Delta W_{qr} &= e_r \cdot A_q = (-1.6) \cdot (1) = -1.6 \\
\Delta W_{ps} &= e_s \cdot A_p = (1.4) \cdot (0) = 0 \\
\Delta W_{qs} &= e_s \cdot A_q = (1.4) \cdot (1) = 1.4
\end{aligned}$$

Here, too, the new weights are the sums of the old weights and the corresponding changes.

$$W_{pr} = \quad 2.5 + 0 = 2.5 \qquad W_{qr} = 1.2 - 1.6 = -0.4$$
$$W_{ps} = -1.4 + 0 = -1.4 \qquad W_{qs} = 2.1 + 1.4 = 3.5$$

The new weights after one backpropagation pass are shown on the righthand side of Figure 9.35.

How do these new weights perform on the same training pair (F, T)/T? Recalling that $A_p = 0$ and $A_q = 1$, we have

$$W_{pr} \cdot A_p + W_{qr} \cdot A_q = (2.5) \cdot (0) + (-0.4) \cdot (1) = -0.4, \text{ so } A_r = 0$$

and

$$W_{ps} \cdot A_p + W_{qs} \cdot A_q = (-1.4) \cdot (0) + (3.5) \cdot (1) = 3.5, \text{ so } A_s = 1$$

and

$$W_{ro} \cdot A_r + W_{so} \cdot A_s = (-0.6) \cdot (0) + (2.4) \cdot (1) = 2.4$$

Hence, now $A_o = 1$, and o returns the value T. That is, backpropagation has adjusted the weights to perform correctly on this input/output pair.

The next step would be to try these weights on each of the other members of the training set, (T, T)/T, (T, F)/F, and (F, F)/T, and use backpropagation to adjust the weights for any pair that the calculation disagrees with. At the end of one run through the entire training set, we rerun the first pair (F, T)/T, because it may be that weight adjustments for the other three pairs no longer satisfy the first pair. If necessary, repeated runs through all four pairs in the training set are executed until all four are satisfied. (What results do you get for each member of the training set with the last weights found in the example?)

For more complicated networks, backpropagation is more intricate, but this example conveys the basic idea.

The process of supervised learning can be viewed as a dynamical system. The state of the system is the network's program — that is, its set of weights. The weight-changing algorithm (for example, backpropagation) defines the dynamic, the system's rule of change. Training continues until no improvements in the output error are obtained. The dynamics has then been attracted to a fixed point. Unfortunately, that fixed point may correspond to a set of weights that produces an unacceptable level of output error. When this is the case, techniques for dislodging the state and pushing it into another, more promising basin of attraction must be implemented. There are several other practical concerns in training networks to perform desired tasks: What is the optimal size of the network? How many hidden layers should be chosen? How is the problem best described? And how many examples should be included in the training set? Bigger may sound better at first, but large networks take longer to train, and better performance can often be obtained with fewer neurons. Also, like humans, neural networks can be pretty good at recognizing categories and structures but pretty bad at making precise calculations. Thus neural networks are more likely to give a satisfactory answer to the question "Will the Market go up tomorrow?" than to the question "By how much?"

Let's investigate some of the peculiarities of artificial neural networks by taking a simple, specific example. We choose the following architecture: two input neurons, one output neuron, and a hidden layer consisting of N neurons that we can specify later. Each input neuron is connected to each hidden-layer neuron, and each hidden-layer neuron is connected to the output neuron. See Figure 9.36 (where the N = 3 case is shown). Real

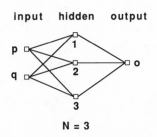

Figure 9.36 *Architecture of an example network.*

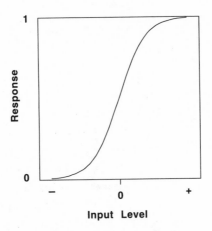

Figure 9.37 *Plot of the firing level of a real neuron versus the input stimulation it is receiving.*

neurons are observed to have graded responses to stimuli. That is, a neuron isn't simply "on" or "off." Its firing level depends in a continuous fashion on the input stimulation it is receiving— something like what is shown in Figure 9.37. Networks with binary rules (on or off), frequently are found in practice to be difficult to train. Often a smoothly continuous activity rule produces better performance. We choose such an activity function for our neurons as well. (In cellular automata language, each neuron has infinitely many states, each of which is represented by a number between 0 and 1.) Finally, we feed error information back into the network in order to adjust weight values according to the backpropagation method (appropriately modified for graded response).

Figure 9.38 shows the results of training the network defined above on a training set representing the logical inclusive OR function. Plotted are the number of times the training set had to be reexamined before acceptable mastery was attained for networks of varying sizes. (Here "acceptable" was taken to mean that the output neuron's activity had to be no greater than 0.1 to be in agreement with "false" and no less than 0.9 to be in agreement with "true.") Usually, though not always, an increase in network size is accompanied by an increase in efficiency of

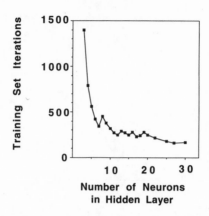

Figure 9.38 *This plot suggests that larger neural networks learn more efficiently than small ones.*

attaining mastery: the bigger the brain, the easier the learning task (usually).

You may wonder why we would want to use 10 or more neurons to perform the same task that three can perform. Of course, three neurons *can* compute simple logical functions. And if all you wanted out of life was to find and store food and to procreate occasionally, you could get away with far fewer neurons than you have, too. Ants do their thing just fine with pretty tiny brains. Simple logic functions and other rudimentary tasks can be computed by very small networks if the operational rules are selected carefully at the outset. It takes a lot more neurons to learn the rules when they aren't supplied by hardwiring, however.

In the supervised training example we have just considered, the network is left to construct its own rules; it has to determine weights of appropriate values for its various connections in order to obtain satisfactory agreement with what it is told are the "correct" answers. Inspecting the weights chosen by a given network to accomplish a particular task yields little understanding of how those choices were made. We would be hard pressed to explain why the learned connections are the way they are. In a sense, learning networks (like humans) construct their own idiosyncratic views of reality through their acquired experiences.

The following exercise illustrates the nonintuitive structure of a trained network. We take a network of size 10 that has been trained to compute inclusive logical OR, and we remove one of the neurons in the hidden layer. Then we ask the "damaged" network to predict the outputs for the inputs of the training set it used to know before the damage was inflicted. In general, its ability to remember the right answers is degraded; the level of degradation depends on which neuron is killed. Subsequently, we retrain the 9 still-functional neurons. The number of examinations of the training set needed to retrain the damaged network is almost always less than the number of examinations it takes a network of size 9 to learn from scratch. Though its first attempt to reconstruct the proper outputs is often very far off the mark, the damaged network *does* retain some recollection of its previous experience. That not every hidden neuron is equally important for computing OR is emphasized by Figure 9.39, where the number of retraining steps is plotted for each of the 10 hidden neurons as the killed neuron. In this particular example, the 10-neuron network could have lost neuron 8 without significant effect on its performance in computing OR.

Several architectures besides the one shown in Figure 9.34 have been studied, and each has its own peculiar attributes— some advantageous, some not. Different network types have

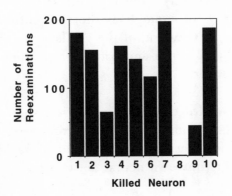

Figure 9.39 *Plot of the time needed to retrain a network with 10 neurons on the training set for inclusive OR after one neuron is removed versus which of the neurons is removed.*

been observed to exhibit fascinating, *emergent* properties (properties that are not programmed in but that emerge somehow from the complex of neuron–neuron interactions). These include the ability to generalize, to identify and categorize common characteristics, and to remember associatively.

Some trained networks can supply at least approximately correct outputs for inputs that, although they are similar to ones on which they were trained, were previously unknown. Such generalization behavior suggests more than mere memorization; it implies a kind of mastery of the concepts relating input to output. The ability to generalize is probably the single most astonishing feature of artificial neural networks. It is also the most practically important. It means that suitably trained networks can be given fresh input data and asked to make judgments about appropriate responses. In some vague sense, such networks "get it." (Recently, a contest to supply the best guess for entries 1001 through 1100 in a noisy time series, given the first 1000 entries, was won by a neural network. The time series was generated by deterministic chaos, and the network *got it*: one dynamical system tuned in to the determinism in the output of another.)

Some networks are particularly adept at classification. Often these networks function by clusters of neurons joining in collective action. Often no supervision is required; classification within a list can emerge simply from repeated exposure of such networks to the list. An example is the recognition, by different neural clusters in a network, of the similarity of the steep-sided letters A, H, M, and N; the similarity of the rounded letters C, G, O, and Q; and so on.

Finally, still other networks display associative memory. Armed with a wealth of stored memories, some networks can rapidly fill in missing information or correct erroneous information. For example, doesn't the input Dnld Dk rapidly conjure up a cartoon bird in your own neural network?

The Evolution of Organization

Though many tens of thousands of scientists actively study how the properties of living matter are governed and implied by the laws of physics, we sometimes hear their pursuit challenged, from other quarters, on the basis of one of the central pillars of physical science, the Second Law of Thermodynamics. The argument is essentially that the existence of highly organized, living structures cannot be explained within the framework of conventional physics because the Second Law forbids those structures. Let's take a moment to trace the origin of this claim and to assess its validity.

How many times have you said to yourself, "My desk (room, check book, life, . . .) is a mess. I need a little organization"? Organization suggests arrangement, interconnection, multiple pieces functioning together as if with a purpose. Organization certainly suggests underlying rules: the presence of order. And, of course, organization implies utility. A messy desktop in total disarray is so much less useful than one that is organized. An incoherent pile of papers, books, computer disks, pencils, and the like impedes information retrieval ("Where the devil are those meeting notes?") and resource use ("Why can't I lay my hands on my red pen?")

The Second Law of Thermodynamics says, roughly, that in equilibrium, any isolated system (a system unable to exchange energy and matter with its surroundings) tends to achieve a state — consistent with the amounts of energy and matter it contains and the space available to it — of maximum disorganization. The Second Law tells us — again, roughly — that as equilibrium is approached, the natural tendency of things is to run downhill and fall apart: energy and matter enter states of decreasing usefulness. Incidentally, disorganization has a quantitative measure called *entropy*. Thus the Second Law is often taken to say that equilibrium for an isolated system corresponds to a state of maximum entropy (consistent with the amounts of energy . . .).

This chapter abounds with examples of "spontaneous appearance of organization" — naturally emerging situations in which energy and matter are concentrated into useful forms. How are these sudden blossomings of utility reconciled with the Second Law of Thermodynamics? Relevant to this question are two crucial words: "equilibrium" and "isolated." The Second Law is about equilibria; it deals with end-states, not with the means by which the end-states are achieved. In addition, the Second Law is about systems that cannot swap energy and matter with their surroundings. All the interesting examples of the abrupt appearance of organization that we see around us (such as the emergence of life on Earth) occur in situations that are far from equilibrium and where energy and matter can be exchanged. These examples are transient, not equilibrium, phenomena.

The Second Law also says "tends." That's because it is a statement of likelihoods, not absolutes. Even in equilibrium, systems can evidence some dynamic variability and still be consistent with the Law. Small fluctuations—local concentrations of energy and matter—ripple throughout such systems, but they quickly peter out, leaving no trace. Far from equilibrium, as all of the self-organizing systems we have explored are, small fluctuations can amplify, leading to large consequences. The work of Ilya Prigogine [124] and colleagues provides a rough guideline to which nonequilibrium amplifications are most likely to occur: processes that create organization in one place typically create *maximum* disorganization (that is, entropy) elsewhere (generally in the form of "waste heat"). Far from equilibrium, energy and matter *can* self-assemble into useful forms in one place, but they do so at the expense of utility in the surroundings. Life is certainly not barred by thermodynamics, but neither is it ensured. Given two possible scenarios that an amplifying fluctuation might follow, the one that is more likely to occur is the one that creates more disorganization. Unfortunately, though this broad principle may be useful in *excluding* some evolutionary scenarios, it does not help us decide which are possible in the first place. For that we will need a different theoretical framework.

Artificial Life

Cellular automata show an amazingly diverse and remarkably lifelike repertoire of behaviors: spontaneous development of spatial structure, self-replication, locomotion, the ability to compute and to learn. Little wonder that they play a central role in the rapidly expanding research program called *Artificial Life*. A-Life is the study of life as a collection of processes, as a kind of logic, regardless of material expression. That is, aluminum and silicon are as capable of manifesting life — in the A-life view — as is carbon. A-Life researchers try to investigate evolution, the genesis of forms, neural processing, emergent properties, and so on in controlled, repeatable situations facilitated by time compression. (Computer-based life forms can metabolize, grow, reproduce, and mutate incredibly quickly.) The inhabitants of A-Life research laboratories are both abstract information structures — automatalike creatures that flock together, perform purposeful tasks, replicate, and evolve — and physically realized entities — robots capable of pattern recognition, motion, and control. At this stage, many examples are whimsical and even cute (the remarkably infectious and sometimes pernicious information structures known as computer viruses notwithstanding), but this toyland quality will surely change as models approaching ever closer to the essence of life are developed. And what we will learn about the greatest mystery of all will surely be wonderful.

The field of Artificial Life was given its name by Chris Langton [125]. His idea was to study the abstract features of evolution and life, and he selected the name to parallel the term *Artificial Intelligence*. Many other people are involved in related work; we sample here but a few of the threads of the A-Life tapestry. The over-arching theme in A-Life studies — as throughout this book — is that nonlinear, iterative processes can generate states of unexpected complexity; you don't have to be careful to program the complexity in, it (sometimes) just emerges.

A first example harks to life's very beginnings: to chemical organization. The standard picture of the origin of life on Earth begins with the spontaneous formation of building blocks of biomolecules in the seas (an idea made compelling by experiments of Harold Urey and Stanley Miller). In this picture, as the availability of these putative building blocks increased, they gave rise to molecules of ever greater complexity and, eventually, to molecules capable of self-replication. Unfortunately, calculation of the time required for sufficiently complicated molecules to assemble by random interactions casts doubt on this hypothesis [126]. Early work by Melvin Calvin and Manfred Eigen, and more recent work by Stuart Kauffman [127], suggest a way around this problem: instead of the larger molecules arising through random encounters, perhaps they might form through series of autocatalytic reactions. That is, one kind of molecule increases the likelihood of a second kind's forming, which in turn speeds the formation of yet a third kind, . . . , and eventually one of these speeds the formation of the first kind. Such reactions are a chemical analogue of feedback and can, in principle, produce what Kauffman calls "order for nothing."

The emergence of global organization from local rules is seen in Craig Wilson's "boids," digital birds [128]. Boids move according to three rules: each boid tries to keep a minimum distance between itself and everything around it, each boid tries to move with the same speed and direction as the other boids near it, and each boid tries to move toward the center of the group of nearby boids. Amazingly, these three natural-sounding rules, which prescribe merely how each boid interacts with those near it, give rise to flocking and graceful avoidance of obstacles.

Emergent order in a different context is observed in the work of John Holland. Applying the notions of adaptation and recombination from biology, Holland developed "classifier systems" and "genetic algorithms" for computer programs [129]. Seeking a program to solve a particular kind of problem, Holland started with some initial distribution of nonsense programs containing functional subunits. The programs were tested on the problem, and each was assigned a "fitness" based on how close it came to solving the problem. The more "fit" programs were mated and produced offspring, which replaced some less fit programs in the initial set. This sexual reproduction of programs was achieved by treating each program as a subunit string, randomly selecting a point on the string, and switching the subunit strings after the chosen point. For example, programs abcdefgh and 12345678 (each character in these strings is meant to represent a functional subunit) might produce "children" abc45678 and 123defgh. Fitnesses were recomputed and the process continued (with occasional mutations allowed). Eventually, the process led to programs that sometimes were remarkably adept at solving the original problem.

(An early application of genetic algorithms was made by David Goldberg [130] in simulating the control of the flow of natural gas through a pipeline. The delay time between (1) instructions to change valve and compressor settings and (2) the change in gas pressure down the pipe makes this problem resistant to analytic solution. Goldberg's system was trained like any human operator is trained. An unexpected development occurred when the system evolved a "default hierarchy" of rules. That is, the first rule controlled the pipeline under the normal conditions, and other rules developed to handle problem situations. None of this was written into the original program, so it is an example of "emergent behavior." Programs have been evolved using genetic algorithms to solve several problems, including the design of communications networks and jet turbines, and they may figure significantly in the design of the next generation of computers.)

Though genetic algorithms may have a profound impact on computer programming, Holland was not completely satisfied with applying fitness from the outside—that is, imposing it via the programmer. In ecosystems, the driving force is not really the pursuit of a "passing grade" but rather coevolution: what each individual is doing relative to all others. Holland constructed a coevolutionary simulation, Echo, in which digital creatures wander around a digital landscape competing with one another for digital resources [131]. Each animal has two "chromosomes," one for attack and one for defense, made of strings of resource symbols. When two animals encounter one another, each compares its attack chromosome with the defense chromosome of the other. The winner eats the loser in the sense that all the stored resources of the loser go into the winner's "reservoir." When an animal accumulates enough resources to make copies of its chromosomes, it produces a child, though there may be some mutations in the chromosome copies that are passed along.

Early in the simulations, animals developed increasingly complex attack and defense chromosomes—an "evolutionary arms race"—and also split into different species. Later, "markers" were added to the chromosomes that allowed individuals to identify others with whom they would prefer to trade than fight. In this version, cooperation between kindred individuals began to appear, but so did lying and mimicry.

Thomas Ray's digital world, called Tierra [132], consists of a "soup" of computer memory in which Ray places some small programs that are capable of self-reproduction. To drive the evolutionary process, two sorts of mutations occur. At a low rate, random program elements are selected and changed. This is analogous to the effect of cosmic rays. The second type of mutation occurs during replication, when instructions are randomly changed at a slightly higher rate than the "cosmic ray" mutations. In addition, instructions have some low, random probability of being executed incorrectly. Several million iterations after this simulation was started running, a new type of program, "parasites," evolved. Parasites are much shorter than the original program, because they do not contain the reproduction instructions. Rather, they run the reproduction instruction of a nearby regular program. Evolutionary value accrues to shorter programs because they can reproduce more rapidly. Even though this is a sort of benign parasitism (the parasite does no direct damage to the host program), eventually "hyperparasites" evolve. The latter programs can reproduce on their own, and when a parasite tries to use a hyperparasite's reproduction instructions, the hyperparasite "reprograms" the parasite to reproduce copies of the hyperparasite. The hyperparasites force the parasites into extinction.

As in Holland's Echo, Tierra exhibited the evolution of liars. Liar programs say they are twice as long as they really are, effectively doubling their reproduction rate. (The amount of computer time allocated to each program is proportional to its reported length.) Once a species of liars appears, all species must become liars in order to survive. Longer Tierra simulations have shown extended periods of quiescent equilibrium, with occasional bursts of intense activity, mimicking the "punctuated equilibrium" model of evolution described by Eldredge and Gould.

Following what Echo revealed, Holland and Brian Arthur, together with others, approached the problem of building a digital stock market [133]. Instead of assuming the perfectly rational agents posited in neoclassical economics, their simulation began with stupid agents who learned from local experience. Even this simple stock market exhibited bubbles and crashes similar to real market activity, contrary to the predictions of neoclassicist theory.

Out of these many simulations, a common feature began to appear. Norman Packard pointed out that natural selection and genetic algorithms drive these systems to the "edge of chaos": states that correspond to Wolfram's class IV automaton in Langton's analysis. Packard's own simulations [134] involved starting with a collection of cellular automata rules, causing them to evolve by means of a genetic algorithm, and assessing their fitness in terms of their ability to perform a particular calculation. In addition, Kauffman and Sonke Johnsen [135] constructed an ecosystem simulation in which coevolution drove the system to the edge of chaos. Kauffman observed that the edge of chaos gives the highest mean fitness for members of the community; it is here that the community adapts most easily to changes and builds the best models of the environment. Too far within the ordered regime and the species are frozen into local fitness states, most of which surely are poor. Too far into the chaotic regime and every change affects the fitness of all others too much: no one has time to adjust to recent changes before others occur. These models suggest that A-Life simulations may provide new insight into how order evolves and into why evolution seems impelled toward greater complexity.

Perhaps Artificial Life will teach us that life is computation and organization, that life is really software emerging from the bottom up, that information is the "vital force" of the alchemists, that evolution pushes us right to the brink of chaos, and that the states of complex adaptive systems (such as the economy) depend sensitively on the paths they take. Perhaps consciousness is an emergent property. Perhaps increasing complexity is a natural consequence of nonlinear processes churning far from thermal equilibrium.

Perhaps the origin of the vast plurality and richness of the world is not forever beyond our understanding.

c h a p t e r

10

Certainly Not the Last Word

FRACTALS, CHAOS, self-organization, emergent properties—the stuff of the study of complexity—have begun to appear as potentially useful tools and guiding principles in many areas of human intellectual endeavor. We conclude this book with a very quick, random (or is it chaotic?) sampling of some of these, noting that for the most part, the ideas discussed here are still in the formative stages of development.

Smelling, Dreaming, Thinking Chaos

Human physiology is a collection of time-varying systems that are interconnected with varying degrees of strength. Growing interest in the role (if any) that deterministic chaos plays in human health and behavior has resulted in a proliferation of conferences with such themes as "The Head and Heart of Chaos" and in the appearance of professional organizations with titles like "Society for Chaos Theory in Psychology." As we have mentioned, study of the heart as a nonlinear dynamical system is a lively research activity that may have profound clinical implications. Investigation of the nonlinear dynamics of the brain is a similarly vigorous pursuit.

Some properties of chaotic dynamics suggest a qualitative similarity to some unexplained psychological behaviors, including the perplexing coexistence of predictability and unpredictability exhibited by individuals and the observation that small causes can lead to large psychological effects. Is there more to this similarity than mere coincidence? The brain and the nervous system constitute a real, "wet" neural network in which electrochemical information flows from sensory receptor sites — input neurons — to the "hidden" processing neurons of the brain and sometimes back out in the form of actions. How that information sloshes around in this network makes us what we are. For illustrative purposes, we give here three examples of psychophysiological studies designed to ferret out deterministic chaos in real neural systems.

Scents. Have you ever recognized a familiar face after just a fleeting glimpse or a familiar voice after hearing only a few words? This common experience, which is termed preattentive perception, may be an example of chaos in the brain — where a small input produces a rapid change in a complex pattern of neural activity. Recent studies by Walter Freeman and his colleagues [136] have led Freeman to speculate that " . . . chaos underlies the ability of the brain to respond flexibly to the outside world and to generate novel activity patterns, including those that are experienced as fresh ideas." [137]

Freeman's work deals with scent recognition in rabbits. Somewhat like the different sensitivity of different taste buds to certain flavors, different scent receptors respond more strongly than others to certain smells. Only a small number of the receptors that are sensitive to a given smell are active at any instant, and the pattern of active receptors, monitored by electroencephalographic (EEG) probes, changes unpredictably from moment to moment. Even though the activity of scent receptors changes constantly, each exposure to a familiar scent produces a particular map of neuron activity on the surface of the olfactory bulb in the rabbits' brains, and this map is a signature of the scent. These maps seem to correspond to attractors of a dynamical system. This view is buttressed by computer simulations, with which Freeman was able to simulate EEG behavior close to what is actually observed. In these simulations, attractors for familiar scents themselves appeared to be chaotic (not simple cycles or fixed points). Scent familiarity is known to be an associative memory system: often only minute stimuli are sufficient to trigger a full-blown scent recognition or at least a characterization of the odor as belonging to a class of similar scents. The variable, non-lock-step nature of chaotic attractors apparently can facilitate the development of such similarity classifications.

Dreams. What does your brain do when you sleep? Generally, five levels of sleeping brain activity are recognized:

1. You drift in and out of sleep.

2. You sleep continuously but can be awakened by a small noise.

3. A larger noise is needed to awaken you.

4. You sleep very deeply and can be awakened only by a major disturbance.

5. You dream.

This last stage is called REM (rapid eye movement) sleep. These different stages of sleep can be distinguished by measuring the small electrical currents in the brain with EEG probes and looking for particular rhythms.

To explore brain dynamics during sleep, Agnes Babloyantz [138] has treated the output of one channel of an EEG as a time series and has used the method of delay plots (see Chapter Six) to try to assess the complexity of the brain state dynamics. She finds dimensions of the delay plots for the data of about 5.03 for stage 2 sleep and about 4.08 for stage 4 sleep. This suggests that the brain activity may be determined by a reasonably low-dimensional dynamical system during these sleep stages. In REM sleep, which is often associated with creative and complex mental processing, the dynamics of the brain seems to be much more complicated: the measurements do not reveal a finite dimension, suggesting high-dimensional dynamics.

Thoughts. In an attempt to link physiological dynamics more directly with mental states, Dana Redington and Steven Reidbord [139] have examined the electrocardiogram (ECG) output of subjects undergoing psychotherapy. During such sessions, subject and therapist engage in conversations that tend to wander over the subject's entire psychological landscape. As the subject's thoughts change, so does the subject's ECG. The noisy heart-beat time series shows varying degrees of complexity that are clarified by making delay plots. The most complex delay plot patterns are correlated with moments when the subject reports being most "therapeutically involved" — that is, most involved with the emotions and thoughts that are central to the session's focus. Suggestively, the complexity of the ECG data is also often greatest when the therapist reports greatest empathy with the patient.

A constant state of chaos in the brain can give a continual supply of novel activity patterns that promote the development of new and useful neuronal interconnections. Chaos may be the tool the brain uses to allow itself the flexibility to analyze and assimilate new data, to engage in problem solving by trial and error, and to generate insight and make creative, intuitive leaps. Studies such as the ones discussed here hold promise for elucidating human thought and for optimizing and maintaining our intellectual potential.

Fractals and Music

In Chapter Six we described Richard Voss's work on fractal music and 1/f noise. Here we briefly mention three other avenues through which fractals and music may be related: direct fractal constructions of music, the Hsüs' work on "musical Cantor Sets," and Schenker's theory of levels in music.

Composing. Figure 10.1 shows an example of the first and second layers of a polyphonic composition based on applying the pattern of a motif to each note of the motif. The motif rules are as follows: (1) Select a first note. (2) Go up two notes and cut the duration in half. (3) Go down one note, making the duration one-half that of the starting note. (4) Go down one note and return to the starting duration. (5) Go down one note, retaining

Figure 10.1   *A self-similar ditty.*

the starting duration. The self-similarity of this type of construction is apparent, though certainly we must use more than two levels to make any claim of a fractal pattern.

Cantoring. Recent work by Kenneth and Andrew Hsü [140] provides another illustration of the fractal nature of music. When appropriate patterns of notes are removed from some compositions (much like removing the middle thirds from an interval to construct a Cantor Set), the resulting "skeletal" version sounds convincingly like another piece by the same composer. For some compositions, the Hsüs observed that the basic patterns persist through this fractal reduction process until only $\frac{1}{64}$ of the original notes remain. This is strong evidence for the persistence of pattern over many lengths in these compositions.

Decomposing. Heinrich Schenker's approach to musical analysis [141] involved most fundamentally a dissection of the work into hierarchically ordered structural levels, these levels being ever more complex elaborations of some basis. In this sense, in the 1920s Schenker posited a decomposition of a piece of music exhibiting the same form over several levels—that is, a fractal aspect of the piece. One feature of Schenker's analysis is the notion of "motivic parallelism," where a motif is repeated over several levels. For example, Beethoven's Sonata in A, Op. 110, exhibits this motivic organization in that much of the piece can be "derived" from the first four measures [142]. Schenker asserted that great music manifests repetition across levels ("concealed repetition"), and yet his writing never provides abstract methodologies for locating such repetitions. Rather, this type of analysis, which is based on the recognition of subtle modulations, is mastered only with arduous study. Perhaps the point is most clearly stated by Schenker himself: "music was destined to reach its culmination in the likeness of itself." [143]

Although rigorously fractal compositions are mechanical and sterile, fractal-like concepts permeate the works of modern composers. The compositions of Philip Glass (for example, "Floe" of *GlassWorks*) often exhibit algorithmic self-similar features. And Gyorgy Ligeti and Charles Wuorinen, for example, deliberately incorporate self-similarity into their compositions: music begun with an idea for a large-scale form, with smaller pieces filled in as reductions (of some sort) of a grand theme.

Fractals and Chaos in Fiction

Fractals and chaos have started cropping up as subjects in literature, and now that we recognize chaos and fractals, we see that they have occurred (though not, perhaps, labeled as such) to authors in the past. Because nature exhibits fractals and chaos, and because literature in some way reflects nature, this "fractal archeology" should come as no surprise. Here is a selection of examples, reflecting nothing more than the peculiarities of our own reading tastes.

One of the main characters of John Updike's *Roger's Version* is a computer science graduate student seeking some fingerprint of God in fractal structures. The text includes discussions of cellular automata, fractal trees, the Koch Curve, and the Mandelbrot Set. The nonmathematical activity of the story, which revolves around two adulterous love affairs, provides a metaphorical parallel to the challenge that chaos poses to the notions of simplicity and complexity.

Arthur C. Clarke's *The Ghost from the Grand Banks*, which deals primarily with efforts to raise the *Titanic*, includes a description of exploring the Mandelbrot Set (both woven into the story and as a technical appendix), a castle with a lake whose shape mirrors the main features of the Mandelbrot Set, and a typical Clarkean twist involving a very peculiar psychological effect of extensive exploration of the Mandelbrot Set.

Michael Crichton's *Jurassic Park* explores the chaotic dynamics of a population of genetically engineered dinosaurs produced for a theme park on an island off the coast of Central America. The story is punctuated by descriptions of the relevance of chaos theory, delivered by one of the characters, a mathematician who is modeling the nonlinearities of the dinosaur ecosystem. The development of the plot offers a memorable example of sensitive dependence on initial conditions.

Kate Wilhelm's *Death Qualified* involves a psychological experiment through which fractal imagery alters the archetypes of perception, opening up vistas both amazing and terrifying. The result is such a surprise that the clues given along the way are (at least for us) twisted to fit into a more familiar reading of reality.

In *Einstein's Dreams*, Alan Lightman describes alternative versions of time that he imagines to have appeared in Einstein's dreams when he was developing the special theory of relativity. In the dream of 16 April, 1905, time flowed like water in a stream, occasionally transporting people through whorls into the past. In a chilling portrait of sensitive dependence on initial conditions, these temporal castaways tried to interact as little as possible with the past in which they found themselves, for fear of having profound influences on the future, their own time.

Stanislaw Lem's *The Invincible* presents a world inhabited by metallic "insects" that, when unfamiliar circumstances arise, self-organize into a vast cloud, a physical realization of a three-dimensional cellular automaton. Although the individual insects are too simple to respond effectively to changing environments, the aggregate cloud—a product of inorganic evolution—can respond to threats with terrifying power and sophistication.

In *The Eye*, Vladimir Nabokov speculates on the effects of changing an event in one's past and following through the cumulative effect of the resulting variations. Nabokov comments specifically on the branching structure of possible lives, anticipating the form of the logistic map bifurcation diagram. Similarly, Jorge Luis Borges's *The Garden of Forking Paths* describes an ancient Chinese novel in which all the bifurcations occur. (Hence the theme of the short story really is time itself, allowing for multiple realities.)

Fractals make a humorous appearance in Flan O'Brien's *The Third Policeman*. Here the "Atomic Theory of Matter" is interpreted in this fashion: everything is made up of small particles of itself—for example, a sheep is made up of little bits of sheepness. In particular, a person who has spent a lot of time riding a bicycle over rough roads has gotten many of her or his atoms interchanged with those of the bicycle. Hence the person becomes more bicyclelike, and the bicycle more humanlike. The person is unable to stand still unless resting an elbow against the wall or a foot against a curbstone. And in cold, wet weather, the bicycle keeps inexplicably appearing in the kitchen near the warm stove. This is especially humorous in view of our discussion of the scale limitations of physical fractals: we usually cite the atomic scale as a certain lower cut-off for the scales on which self-similarity is exhibited.

Finally, as a sign that fractals have caught the eye of the general public, an episode of the CBS television show *Murphy Brown* dealt with interviewing a physicist (played by Buck Henry) engaged in work with fractals.

Fractals and Chaos in Literary Analysis

Construction. A goal of certain types of writing is to act as a window on another world — sometimes a realistic world, sometime a metaphorical world. This is a difficult task, especially for poetry, which usually has only a small number of words with which to create this illusion. Lucy Pollard–Gott [144] has suggested that through fractal patterns of word or sound repetitions, the ear sets up a regression that the mind continues, supporting the illusion that the poem's few words sketch a picture of a whole world. Pollard–Gott uses poems of Wallace Stevens to support this claim. Imagine a line made up of little boxes, each box corresponding to a word of the poem, left to right as the poem is read. For example, the nursery rhyme

Mary had a little lamb,
Little lamb,
Little lamb,
Mary had a little lamb,
It's fleece was white as snow.

would be represented in this way by the line of boxes

The left-most box corresponds to *Mary*, the next-to-the-left-most to *had*, . . . , the right-most to *snow*. In this arrangement, the word *lamb* corresponds to the filled-in boxes in this graph:

This is hardly a fractal pattern, but then this is hardly great poetry. Performing a similar analysis using the word *know* in Stevens's poem "The Sail of Ulysses (Canto 1)," Pollard–Gott observes a pattern of spacings like the third stage in the construction of a Cantor Set. Not exactly that, of course, but just as physical fractals resemble their mathematical cousins only approximately, word patterns in poems are only approximately like a stage in the construction of a Cantor Set.

Is this a fluke? Probably not, because Pollard–Gott produces other examples. Did Stevens consciously construct his poems around Cantor Sets? Again, probably not, but in a sense this is more interesting than if he had done so. What should we conclude from the fact that this great poet sometimes constructed his poems so that patterns of sound repetitions have fractal structures? Perhaps we can conclude that these patterns sound natural and that (much like the repetitious echo patterns in a large cave) they suggest a wide world beyond the little light of our lantern.

Try a similar analysis with your favorite poet. What do you get? Perhaps you might look for patterns of similar ideas instead of similar sounds.

Deconstruction. Postmodern literary criticism thrives on a tension between the reader and the author, and it ultimately relegates the author to a role of secondary importance. After all, the argument goes, the image a text produces for a given reader has at least as much to do with the context of the reader's life experiences as with the impressions the author intended to communicate. Challenging the creative/passive dichotomy of artist/observer, postmodernism applauds the complex, the multi-leveled, the contingent, and the nonlinear. What is perhaps the pinnacle of postmodern thought, deconstruction, inverts the usual hierarchy of writer/reader and then repeats the inversion again and again. Recently, some authors have drawn interesting comparisons between postmodernism and chaos theory. Here we mention two of them: N. Katherine Hayles and Alexander J. Argyros.

Hayles [145] draws parallels between chaos and postmodernism, especially deconstructionism. Sensitive dependence on initial conditions, the source of the unpredictability of chaos, derives from the nonlinearity of the system. In deconstruction, the role of marginal, seemingly trivial elements is elevated to central importance through the nonlinearity of that process. Both chaos and deconstruction rely on iteration—chaos by iterating the defining function or functions, deconstruction through its endless succession of inversions. Deconstruction, like other postmodern theories, celebrates the necessity of unpredictability by asserting that the reversals of deconstructionist analysis undermine our ability to locate the origin in an argument. This is analogous to the unpredictability of chaos resulting from sensitivity to initial conditions and from our inability to specify these initial conditions precisely. Hayles even suggests that postmodernism and chaos may exhibit more than metaphorical parallels—that these disciplines have interesting things to say to one another.

A different point of view is advanced by Argyros [146]. Arguing against deconstruction's infinite inversions of signification and ultimate abandonment of progress, Argyros views human work as a chaotic system and uses this notion to espouse optimism about the possibility of individuals (small causes) producing noticeable changes in the world (large effects). This may be the most forward-looking interpretation of sensitive dependence on initial conditions proposed yet. In spite of Argyros's "call to action," it is not clear that the methods of controlling chaos in engineering can be implemented in literature. On the other hand, neither is it clearly impossible.

Fractals and Art

Graphic Arts. Postmodernism is not restricted to literature and philosophy. For example, it has found expression in graphic design, representing evolution beyond the strict Bauhaus dictum that communication through images should be unambiguous and immediately interpretable. Postmodernist graphic design experiments with multiple layers, using complex images to insist that the reader be aware of the process of reading, and using ambiguity of form to emphasize the role of the reader in constructing the meaning of the work. Here are some instances.

In *Cranbrook Design: The New Discourse* [147], the role of the reader as constructor of meaning in design pieces is explored through postmodernism's fragmentation of form. For example, the essay by Katherine and Michael McCoy is presented in a two-column format that one reads across the gap between the columns, forcing awareness of the process of reading. Both in content and in style, the modernist tradition of a clean distinction between author and reader is challenged here, allowing for multiple readings via the corollary sensitive dependence of meaning on the reader.

Alice Greiman designed an issue of *Design Quarterly* that folded out into a single poster 2 feet by 6 feet in size. Covered with images of many forms, the poster is a rich aggregate inviting multiple viewings and interpretations. In *Hybrid Imagery* [148], Greiman expresses her dissatisfaction with the conflict of dualities through the language of fractals and chaos. The evident chaos of the *Design Quarterly* poster does not act in opposition to any underlying order and meaning it encompasses. Rather, it forces the viewer to create levels of meaning.

The New Museum of Contemporary Art catalogue *Strange Attractors: Signs of Chaos* [149] was designed to reflect the ideas of chaos (a topic of many of the essays), and the reading sometimes is challenging. Essays are printed in different typefaces and on pages of different sizes, and they wander throughout the text. Some essays (for example, Martin Meisel's "Chaos déjà-vu") have increasingly large spaces inserted between words; the pleasant image of light and dark formed by viewing these pages through squinted eyes perhaps balances the difficulty of reading the text. The back cover of the catalogue has two sets of text overwritten at right angles.

The chaotic influence of multiple layering and meanings achieves a total triumph over the message of the text in *Semiotext(e) Architecture* [150]. In many places, the multiple layers are so densely overwritten that the reader has considerable difficulty unpacking the words. For example, the introduction is overwritten with pictures and short messages (such as "Welcome to the ugliest building in the world"), but underneath these one finds (eventually) the manifesto that design and writing can take roles

other than the traditional and, later, the statement that the text
is not a book but rather "a fractal and allegory dispersion and
proliferation" of architecture and theory. The role of design in
communication is emphasized through passages with curved
margins, text and pictures overwritten, and multiplexed texts
(including descriptions of the role of chaos in science). Here the
task of chaos is to invert the roles of text and design in communication and thus to communicate at a different level.

An interesting experiment in self-organization of design was
begun in 1992 by the manufacturer Gilbert Paper. The project
involved seven designers, one of whom created an initial image
and sent it on disk to another of the seven. As the disk circulated
among the experimenters, each took the current design as a
starting point and modified it as little or as much as he or she
chose. As might be expected, the result was multilayered, filled
with ambiguous readings, but interestingly, even those parts that
had been completely removed by subsequent work left their
echoes as initial conditions for what came after. Though complex
and demanding, the final form can be viewed as arising from a
sort of nonlinear iterative processing.

The goal toward which these examples are pointing is that
with multiply layered designs, meaning no longer is transferred
in a predictable Bauhaus style from author to reader. Instead, it
evolves in an unpredictable way from the reader's interaction
with the layers. Even so, of course, the reading is not random, for
the hand of the designer is present throughout. The relationship
of postmodern graphic design to chaos theory is not simply a
metaphor: some designers deliberately use computer-generated
fractal images and chaotic processes in their work. Yet perhaps
the most interesting influence of chaos on design is in the
insistence that the reader participate actively in the assignment
of meaning. Fractal geometry draws its power from insisting on
the primacy of process over form. Postmodern graphic art views
the designer not as the assembler of a static picture but as the
architect of a process—the reading of the design by many
different people.

Sculpture and Gender. In her essay "Chaos Theory and Fractal Geometry: Their Potential Impact on the Future of Art,"

sculptor Rhonda Roland Shearer argues that the major innovative periods in Western art — the Renaissance and modern art — coincided with the appearance of new geometrical models [151]. Renaissance art was bound up with the advent of perspective geometry, whereas the birth of modern art was nourished by the development of non-Euclidean and higher-dimensional geometries at the end of the nineteenth century. Because much of art is concerned with the construction and division of space and form, geometry provides a cultural context of familiar spatial structures. Indeed, many of the artists who participated in the development of modern art were influenced more or less directly by the nineteenth century's geometrical revolution. In *The Fourth Dimension and Non-Euclidean Geometry in Modern Art* [152], Linda Henderson speculates that artists are attracted to these changes in geometry because such changes represent complete breaks with traditional views. Accordingly, they free the artist from old patterns of thinking and offer a new way to approach space. The development of fractals and chaos represents another revolution in geometry, one that Shearer suggests may signal a new stage of innovation in art. Like perspective and non-Euclidean geometries before it, fractal geometry provides a new view of the world.

That fractal geometry transcends the classical dichotomies of "natural" and "artificial" shapes is of particular interest. We have seen — in iterated function systems, for example — that natural-looking shapes (tree, fern, leaf) can be constructed by the "artificial" act of iterating mathematical functions on a computer. This synthesis is achieved only through the use of fractals: natural shapes defy a simple classical geometrical description. Fractal geometry calls into question the validity of categorizing art as either realistic or abstract. In her own works (such as *Pangea* [see Figure 10.2] and *Ambiguous Figures*), Shearer combines both "real" plant forms and "abstract" constructs from classical geometry. The detailed plant aspects and the interactions between plants and abstract geometry of many different sizes lend Shearer's sculptures a sort of fractal scale independence; they are visually engaging structures on many levels. Shearer notes that in merging the natural with the artificial, fractal geometry removes

Figure 10.2 *A bronze work by Rhonda Roland Shearer. "Pangea," 1990, which was installed to commemorate Earth Day 1990 in New York City.*

the distinction between linear-analytic-reductionist-male and nonlinear-intuitive-holistic-female.

The potential of fractal geometry to blur some culturally imposed gender distinctions may have far-ranging consequences. We wonder, for example, what impact this unifying potential may have on diminishing gender bias in mathematics and science education.

What's next? Who knows. The new ideas we've examined in this book are probably applicable in many situations, and certainly *not* in many others. So be a little skeptical. But now is assuredly a time for soaring imagination. The study of complexity in all its many forms is exploding. In a few years, things will have calmed down and the field will have matured a bit. For now, we get to see — and perhaps contribute to — the reckless activity of its youth.

References and Notes

1. Benoit B. Mandelbrot. 1983. *The Fractal Geometry of Nature*. New York: W. H. Freeman and Company. Following all the details of this book requires sophisticated background in mathematics and physics, but as is common with great literature, it can be understood on many levels. The study of fractals has been shaped largely through Mandelbrot's research, and this book is a good window into how the field developed.

2. James Gleick. 1987. *Chaos: Making a New Science*. New York: Viking. This probably is the best-known popular book about chaos. Other good choices are cited in references 3–7.

3. Ian Stewart. 1989. *Does God Play Dice?* New York: Basil Blackwell.

4. Hans Lauwerier. 1991. *Fractals: Endlessly Repeated Geometrical Figures*. Princeton, NJ: Princeton University Press.

5. Ivars Peterson. 1988. *The Mathematical Tourist: Snapshots of Modern Mathematics*. New York: W. H. Freeman and Company.

6. Ivars Peterson. 1990. *Islands of Truth: A Mathematical Mystery Tour*. New York: W. H. Freeman and Company.

7. John Briggs and F. David Peat. 1989. *Turbulent Mirror: An Illustrated Guide to Chaos Theory and the Science of Wonder*. New York: Harper and Row.

8. Ivar Ekeland. 1988. *Mathematics and the Unexpected*. Chicago: University of Chicago Press. A popular book about chaos for more mathematically oriented readers. See Reference 9 as well.

9. David Ruelle. 1991. *Chance and Chaos*. Princeton: Princeton University Press.

377

10. Robert L. Devaney. 1990. *Chaos, Fractals, and Dynamics: Computer Experiments in Mathematics.* Menlo Park, CA: Addison-Wesley. This book is among the simplest introductory texts, requiring only high school algebra.

11. James Sandefur. 1990. *Discrete Dynamical Systems: Theory and Applications.* Oxford: Oxford University Press. This is an introductory text for those with a good background in calculus.

12. Robert L. Devaney. 1992. *A First Course in Chaotic Dynamical Systems: Theory and Experiment.* Reading, MA: Addison-Wesley. An introductory text that is accessible to readers with good calculus backgrounds. See Reference 13 as well.

13. Denny Gulick. 1992. *Encounters with Chaos.* New York: McGraw-Hill.

14. Heinz-Otto Peitgen, Hartmut Jürgens, and Dietmar Saupe. 1992. *Chaos and Fractals: New Frontiers in Science.* New York: Springer-Verlag. This introductory text covers a considerable amount of material without many formal prerequisites, though the reader does need some practice in mathematical thinking.

15. G. L. Baker and J. P. Gollub. 1990. *Chaotic Dynamics: An Introduction.* Cambridge: Cambridge University Press. A text for those with a solid physics background. See also texts cited in references 16–20.

16. N. Tufillaro, T. Abbott, and J. Reilly. 1992. *An Experimental Approach to Nonlinear Dynamics and Chaos.* Redwood City, CA: Addison-Wesley.

17. Francis Moon. 1992. *Chaotic and Fractal Dynamics: An Introduction for Applied Scientists and Engineers.* New York: John Wiley & Sons.

18. Edward Ott. 1993. *Chaos in Dynamical Systems.* Cambridge: Cambridge University Press.

19. Robert Hilborn. 1993. *Chaos and Nonlinear Dynamics: An Introduction for Scientists and Engineers.* Oxford: Oxford University Press.

20. Manfred Schroeder. 1991. *Fractals, Chaos, Power Laws: Minutes from an Infinite Paradise.* New York: W. H. Freeman and Company. This text is at the junior-senior level in mathematics. Also see texts cited in references 21–23.

21. Robert L. Devaney. 1989. *An Introduction to Chaotic Dynamical Systems,* 2nd ed. Redwood City, CA: Addison-Wesley.

22. Gerald Edgar, 1990. *Measure, Topology, and Fractal Geometry.* New York: Springer-Verlag.

23. Michael F. Barnsley. 1988. *Fractals Everywhere.* New York: Academic Press.

24. S. Neil Rasband. 1990. *Chaotic Dynamics of Nonlinear Systems*. New York: John Wiley & Sons. A text at the junior-senior level in mathematics for students with some physics background. Also see Reference 25.

25. P. G. Drazin. 1992. *Nonlinear Systems*. Cambridge: Cambridge University Press.

26. Kenneth Falconer. 1990. *Fractal Geometry: Mathematical Foundations and Applications*. New York: John Wiley & Sons. At the mathematics graduate level, this is the most mathematically complete treatment of the topics in this book. This is also true of Falconer's book cited in Reference 27.

27. Kenneth Falconer. 1985. *The Geometry of Fractal Sets*. Cambridge: Cambridge University Press.

28. Jens Feder. 1988. *Fractals*. New York: Plenum. This book is at the physics graduate level.

29. William Poundstone. 1985. *The Recursive Universe: Cosmic Complexity and the Limits of Scientific Knowledge*. Chicago: Contemporary Books. This is a popular book that discusses cellular automata, complexity theory, and artificial life, as are the books cited in references 30–32.

30. M. Mitchell Waldrop. 1992. *Complexity: the Emerging Science at the Edge of Order and Chaos*. New York: Simon & Schuster.

31. Roger Lewin. 1992. *Complexity: Life at the Edge of Chaos*. New York: Macmillan.

32. Steven Levy. 1992. *Artificial Life: The Quest for a New Creation*. New York: Pantheon Books.

33. Eliot Porter and James Gleick. 1990. *Nature's Chaos*. New York: Viking. This is among the many picture books on fractals and chaos; see references 34 and 35 for other good choices.

34. Michael McGuire. 1991. *An Eye for Fractals*. Redwood City, CA: Addison-Wesley.

35. Etienne Guyon and H. Eugene Stanley. 1991. *Fractal Forms*. Amsterdam: Elsevier.

36. H.-O. Peitgen and P. H. Richter. 1986. *The Beauty of Fractals: Images of Complex Dynamical Systems*. Berlin: Springer-Verlag. Though the mathematics is quite advanced, this book contains some of the first widely distributed color pictures of zooms into the Mandelbrot Set.

37. Stephen Kellert. 1993. *In the Wake of Chaos*. Chicago: University of Chicago Press. Relations between chaos, fractals, and the humanities are explored in this book; see references 38 and 39 as well.

38. N. Katherine Hayles, ed. 1991. *Chaos and Order: Complex Dynamics in Literature and Science*. Chicago: University of Chicago Press.

39. Alexander J. Argyros. 1991. *A Blessed Rage for Order: Deconstruction, Evolution, and Chaos*. Ann Arbor: University of Michigan Press.

40. J. Crutchfield, D. Farmer, and N. Packard. 1986. "Chaos." *Scientific American*, December. Over the years, *Scientific American* has published many articles on the topics in this book. Well-written and at a reasonably accessible level, these include this article and those cited in references 41–53.

41. L. Sander. 1987. "Fractal Growth." *Scientific American*. January.

42. J. Fricke. 1988. "Aerogels." *Scientific American*. May.

43. J. Ottino. 1989. "The mixing of fluids." *Scientific American*. January.

44. A. L. Goldberger, D. Rigney, and B. West. 1990. "Chaos and fractals in human physiology." *Scientific American*. February.

45. H. Jürgens, H.-O. Peitgen, and D. Saupe. 1990. "The language of fractals." *Scientific American*. August.

46. P. Bak and K. Chen. 1991. "Self-organized criticality." *Scientific American*. January.

47. Walter Freeman. 1991. "The physiology of perception." *Scientific American*. February.

48. Stuart Kauffman. 1991. "Antichaos and adaptation." *Scientific American*. August.

49. M. Gutzwiller. 1992. "Quantum chaos." *Scientific American*. January. Another perspective can be found in J. Ford, G. Mantica, and G. H. Ristow. 1990. "The Arnol'd cat: Failure of the correspondence principle." in *Chaos/Xaos*. Edited by D. K. Campbell. New York: American Institute of Physics. 477–493.

50. John Holland. 1992. "Genetic algorithms." *Scientific American*. July.

51. G. Hinton. 1992. "How neural networks learn from experience." *Scientific American*. September.

52. R. Ruethen. 1993. "Adapting to complexity." *Scientific American*. January.

53. William L. Ditto and Louis M. Pecora. 1993. "Mastering chaos." *Scientific American*. August.

54. Martin Gardner. 1970. "Mathematical Games." *Scientific American*. October. The Mathematical Games column (including its several later incarnations: Metamagical Themas, Computer

Recreations, or Mathematical Recreations) has described many aspects of fractals, chaos, and related topics. This is among the best-known of these, as are those cited in references 55 and 56.

55. Martin Gardner. 1978. "Mathematical Games." *Scientific American.* April.

56. A. K. Dewdney. 1985. "Computer Recreations." *Scientific American.* August.

57. Jeanne McDermott. 1983. "Fractals will help to make order out of chaos." *Smithsonian.* December.

58. James Gleick. 1987. "Using order to make chaos." *Smithsonian.* December.

CHAPTER ONE

59. Helen Gardner. 1980. *Art Through the Ages*, 7th ed. New York: Harcourt Brace Jovanovich. page 783.

60. Page 1 of reference [1].

61. Pages 19–22 of reference [14].

62. Pages 196–7 of reference [13].

63. H. Joel Jeffrey. 1992. "Chaos game visualization of sequences." *Computers & Graphics* 16, 25–33.

64. Many aspects of which are described in reference [1].

65. M. Barnsley, V. Ervin, D. Hardin and J. Lancaster. 1985. "Solution of an inverse problem for fractals and other sets." *Proceedings of the National Academy of Sciences USA* 83, 1975–77. See also M. Barnsley and A. Sloan. 1988. "A better way to compress images." *Byte* January. 215–233.

66. Daniel Robbins. 1990. *Larry Poons: Paintings 1963–1990.* New York: Salander-O'Reilly Galleries.

67. R. F. Voss and J. Clarke. 1975. "1/f noise in music and speech." *Nature* 258, 317–318.

CHAPTER TWO

68. John Hutchinson. 1981. "Fractals and self-similarity." *Indiana University Journal of Mathematics* 30, 713–747.

69. See reference [23]. Also see Michael F. Barnsley and Lyman B. Hurd. 1993. *Fractal Image Compression.* AK Peters Ltd. MA: Wellesley.

70. Page 96 of reference [23].

CHAPTER THREE

71. Carl A. Carlson. 1991. "Spatial distribution of ore deposits." Geology 19, 111–114.
72. Bernard Sapoval. 1989. "Experimental observation of local modes in fractal drums." *Physica D* 38, 296–298.
73. W. Engelberts and L. Klinkenberg. 1951. "Laboratory experiments on the displacement of oil by water from packs of granular material." *Third World Petrology Congress Proceedings*, 544–554. See also references [28] and [41].
74. Bennett Davis. "Dust demon." *Discover* (March, 1992), 114–116.
75. Heinz-Otto Peitgen and Dietmar Saupe, eds. 1988. *The Science of Fractal Images*. New York: Springer-Verlag, pages 22–35.

CHAPTER FOUR

76. Steven C. Frautschi, Richard P. Olenick, Tom M. Apostol, and David Goodstein. 1986. *The Mechanical Universe*. Cambridge, England: Cambridge University Press. Page 309.
77. Philip Falconer. The Weather Information Service, East Greenbush, NY.
78. Elizabeth Wood, private communication.
79. Lou-Anne Beauregard, private communication.
80. Pages 114–121 of reference [12].
81. Edward Lorenz. 1963. "Deterministic non-periodic flow." *Journal of Atmospheric Science* 20, 130–141.
82. Henri Poincaré. 1892–1899. *Les Methodes nouvelles de la mechanique celeste*. English translation *New Methods of Celestial Mechanics*. 1993. American Institute of Physics.
83. G. Birkhoff. 1927. *Dynamical Systems*. Providence, RI: American Mathematical Society.
84. M. Cartwright and J. Littlewood. 1945. "On nonlinear differential equations of the second order, I: The equation $\ddot{y} + k(1 - y^2)\dot{y} + y = b\lambda k\cos(\lambda t + a)$, k large." *Journal of the London Mathematical Society* 20, 180–189.
85. Stephen Smale. 1967. Differentiable Dynamical Systems. *Bulletin of the American Mathematical Society* 73, 747–817.
86. T. Li and James Yorke. 1975. "Period three implies chaos." *American Mathematical Monthly* 82, 985–992.

CHAPTER FIVE

87. Mitchell Feigenbaum. 1978. Quantitative universality for a class of nonlinear transformations." *Journal of Statistical Physics* 19, 25–52.

88. P. Linsay. 1981. "Period doubling and chaotic behavior in a driven, anharmonic oscillator." *Physical Review Letters* 47, 1349–1352. See also J. Testa, J. Perez, and C. Jeffries. 1982. "Evidence for universal chaotic behavior of a driven nonlinear oscillator." *Physical Review Letters* 48, 714–717.

89. Albert Libchaber. 1982. "Convection and turbulence in liquid helium I," *Physica B* 109 and 110, 1583–1589.

90. Michael R. Guevara, Leon Glass, and Alvin Schrier. 1981. "Phase locking, period-doubling bifurcations, and irregular dynamics in periodically stimulated cardiac cells." *Science* 214, 1350.

91. R. G. Harrison and D. J. Biswas. 1986. "Chaos in light." *Nature* 321, 394–401.

92. R. Simoyi, A. Wolf, and H. Swinney. 1982. "One-dimensional dynamics in a multi-component chemical reaction." *Physical Review Letters* 49, 245–248.

93. R. Shaw. 1984. *The Dripping Faucet as a Model Chaotic System.* Santa Cruz, CA: Ariel Press.

94. C. Maganza, R. Causse, and F. Laloe. 1986. "Bifurcations, period-doubling and chaos in clarinetlike systems." *Euophysics Letters* 1, 295–304.

95. A. Saperstein. 1984. "Chaos—a model for the outbreak of war." *Nature* 309, 303–305.

CHAPTER SIX

96. Michel Hénon. 1976. "A two-dimensional mapping with a strange attractor." *Communications in Mathematical Physics* 50, 69–77.

97. Gabriel Mindlin and Robert Gilmore. 1992. "Topological analysis and synthesis of chaotic time series." *Physica D* 58, 229–242. See also Claire Gilmore. 1993. "A new approach to testing for chaos, with applications in finance and economics." International Journal of Bifurcation and Chaos 3, 581–587.

98. Page 28 of reference [55].

99. Daniel Berlyne. 1971. *Aesthetics and Sociobiology.* New York: Appleton-Century-Crofts.

100. David Berreby. 1993. "Chaos hits Wall Street." *Discover*, March, 76–84.

101. W. Ditto, S. Rauseo and M. Spano. 1990. "Experimental control of chaos." *Physical Review Letters* 65, 3211–3214.

102. E. Ott, C. Grebogi, J. Yorke. 1990. "Controlling chaos." *Physical Review Letters* 64, 2296–2299, 2837.

103. E. R. Hunt. 1991. "Stabilizing high-period orbits in a chaotic system: the diode resonator." *Physical Review Letters* 67, 1953–55; R. Roy, T. W. Murphy, Jr., T. D. Maier, Z. Gills, and E. R. Hunt. 1992. "Dynamical control of a chaotic laser: experimental stabilization of a globally coupled system." *Physical Review Letters* 68, 1259–1262; A. Garfinkel, M. Spano, W. Ditto, and J. Weiss. 1992. "Controlling cardiac chaos." *Science* 257, 1230–1235.

CHAPTER SEVEN

104. Also the "Computer Recreations" column of that issue, reference [56].

105. Bodil Branner. "The Mandelbrot Set." pages 75–105 in *Chaos and Fractals, the Mathematics Behind the Computer Graphics*, ed. R. Devaney and L. Keen. Providence, RI: American Mathematics Society. See pages 80–81.

106. Lavaurs' algorithm: see pages 98–100 of reference [105].

107. A theorem of Douady and Hubbard: see pages 90–91 of reference [105].

108. M. Shishikura. "The Hausdorff dimension of the boundary of the Mandelbrot set and Julia sets." SUNY Stony Brook, Institute for Mathematical Sciences, Preprint 1991#7.

109. See page 80 of reference [105].

110. Hubert Cremer sketched a rough picture of a Julia set in "Über die Iteration rationaler Funktionen." 1925. *Jahresbuch der Deutschen Mathematischen Vereinigung.* 33, 185–210. According to Peitgen, Jürgens, and Saupe (page 123 of reference [14]), this is the first picture of a Julia set.

111. Tan Lei's theorem, see pages 101–102 of reference [105].

CHAPTER EIGHT

112. The original source is A. Cayley. 1879. "The Newton-Fourier imaginary problem." *American Journal of Mathematics* 2, 97. See

also H. O. Peitgen, D. Saupe, and F. V. Haeseler. 1984. "Cayley's problem and Julia sets." *Mathematical Intelligencer* 6, 11–20.

113. J. Curry, L. Garnett, D. Sullivan. 1983. "On the iteration of rational functions: computer experiments with Newton's method." *Communications in Mathematical Physics* 91, 267–277.

CHAPTER NINE

114. J. von Neumann. 1966. *Theory of Self-Reproducing Automata*, ed. A. Burks. Urbana: University of Illinois Press.

115. Wentian Li. 1989. "Complex patterns generated by next nearest neighbors cellular automata." *Computers & Graphics* 13, 531–537.

116. Reference [54].

117. Reference [29].

118. E. Berlekamp, J. Conway, and R. Guy. 1982. *Winning Ways (for Your Mathematical Plays)*. New York: Academic Press.

119. S. Wolfram. 1984. "Universality and complexity in cellular automata." *Physica D* 10, 1–35.

120. C. Langton. 1986. "Studying artificial life with cellular automata." *Physica D* 22, 120–149.

121. Reference [46].

122. P. Bak, C. Tang, and K. Wiesenfeld. 1988. "Self-organized criticality." *Physical Review* A 38, 364–374.

123. William F. Allman. 1990. *Apprentices of Wonder: Inside the Neural Network Revolution*. New York: Bantum Press.

124. Ilya Prigogine and Isabelle Stengers. 1984. *Order out of Chaos: Man's New Dialogue with Nature*. Toronto: Bantam Books.

125. References [30], [31], and [32].

126. Page 122 of reference [30].

127. Pages 124–5 and 128 of reference [30].

128. Pages 241–2 of reference [30].

129. Pages 188–90 of reference [30]. See also reference [50].

130. Pages 191–3 of reference [30].

131. Page 261 of reference [30].

132. Pages 94–104 of reference [31].

133. Page 269 of reference [30].

134. Page 303 of reference [30].

135. Pages 309–12 of reference [30]. See also reference [48].

CHAPTER TEN

136.   Reference [47].

137.   Page 78 of reference [47].

138.   A. Babloyantz. 1990. "Chaotic dynamics in brain activity." *Chaos in Brain Function*, E. Basar. Berlin: Springer-Verlag.

139.   Daniel Pendick. 1993. "Chaos of the Mind." *Science News* 143, February, 138–139.

140.   Kenneth Hsü and Andrew Hsü. 1990. "Fractal geometry of music." *Proceedings of the National Academy of Sciences USA* 87, 938–941. Kenneth Hsü and Andrew Hsü. 1991. "Self-similarity of the '1/f noise' called music." *Proceedings of the National Academy of Sciences USA* 88, 3507–3509.

141.   David Beach, ed. 1983. *Aspects of Schenkerian Theory*. New Haven: Yale University Press.

142.   Page 31 of reference [141].

143.   Page 39 of reference [141].

144.   Lucy Pollard-Gott. 1986. "Fractal repetition in the poetry of Wallace Stevens." *Language and Style* 19, 233–249.

145.   N. Katherine Hayles. 1991. "Complex dynamics in literature and science." Pages 1–33 of ed. N. Katherine Hayles, *Chaos and Order: Complex Dynamics in Literature and Science*. Chicago: University of Chicago Press.

146.   Alexander J. Argyros. 1991. *A Blessed Rage for Order: Deconstruction, Evolution, and Chaos*. Ann Arbor, MI: University of Michigan Press.

147.   Aldersey-Williams, et al. 1990. *Cranbrook Design: the New Discourse*. New York: Rizzoli International Publishers.

148.   Alice Greiman. 1990. *Hybrid Imagery*. New York: Watson-Guptil Publications.

149.   *Strange Attractors: Signs of Chaos*. New York: New Museum of Contemporary Art.

150.   Hraztan Zeitlian, ed. 1992. *Semiotext(e) Architecture*. California: Hraztan Zeitlian.

151.   Rhonda Roland Shearer. 1992. "Chaos Theory and Fractal Geometry: Their Potential Impact on the Future of Art." *Leonardo* 25, 143–152.

152.   Linda Henderson. *The Fourth Dimension and Non-Euclidean Geometry in Modern Art*. Princeton, NJ: Princeton University Press.

ADDITIONAL RESOURCES

Several good videotapes on chaos and fractals provide animated excursions into FractalSpace. Among these are the following: *Focus on Fractals* and *Mandelbrot Sets and Julia Sets*, Homer Smith. *Fractals: an Animated Discussion*, H.-O. Peitgen, H. Jurgens, D. Saupe, and C. Zahlten. *Chaos, Fractals, and Dynamics: Computer Experiments in Mathematics*, Robert Devaney. *Fractals and Chaos in Simple Physical Systems, as Revealed by the Computer*, Frank Varosi and James A. Yorke. *The Beauty and Complexity of the Mandelbrot Set*, John Hubbard.

A considerable array of software gives users hands-on experience, of inestimable value in learning this field. In addition to the package supporting the text, a few of those we have seen include the following: *A First Course in Chaotic Dynamical Systems Software*, James Georges, Del Johnson, Robert Devaney. *The Beauty of Fractals Lab*, H. Jurgens, H.-O. Peitgen, D. Saupe, T. Eberhardt, M. Parmet; *The Desktop Fractal Design System*, Michael Barnsley; *FractInt*, Timothy Wegner, Mark Peterson.

Explorations and Challenges

Extended problem sets separated by Chapter and Section and identified by type (analytic, numerical, computer-based, experimental,
and essay), computer software useful for exploration, laboratory
instructions, and course syllabi are available from the authors for
the cost of mailing and handling. Please write to David Peak, Physics
Department, Union College, Schenectady, NY 12308, or send for
details by e-mail to PeakD@gar.union.edu.

CHAPTER ONE

1.1. Here's a nice way to demonstrate that the starting point in
the Chaos Game is irrelevant. It requires access to a bit of
equipment and is best done as a group project. You will need a
reproducing machine, some indelible markers, some overhead
projection foils, and enough dice and rulers to go around to
everyone in the group. Mark three points A, B, and C on a master
foil. Run off a number of copies on foil blanks. Distribute a copy
to each member of the group. Have each member play the Chaos
Game (Three Corners, One-half Version) on his or her foil, but in
each case *do not* record the first few new points. For example,
have each player mark the first few (about 10) points with a light
pencil that can be erased. Each person should play the game
about 30 times, making about 20 indelible points in each case. Of
course, each should be encouraged to pick a different starting

point. Now collect all of the foils and superpose them so that all the A's, B's, and C's, respectively, exactly overlap. The superposition of indelible marks from the entire set of games should form a piece of the Gasket. The more players, the more nearly complete will be the superposition. Try it. You'll be amazed.

CHAPTER TWO

2.1. Find three transformations that will make an equilateral Gasket with side length 2. (*Hint*: Refer to Figure 2.4.)

2.2. Consider the IFS defined by the following rules:

> Scale by 0.333.
>
> Scale by 0.333, then translate by +0.5 in the x-direction.
>
> Scale by 0.333, then translate by +0.5 in the y-direction.
>
> Collage the pieces.

Starting from the triangle with vertices (0, 0), (1, 0), and (0, 1), observe that with each successive iteration, the total size of the picture shrinks a bit. Show that the base and height of the final Gasket dust are both 3/4. (*Hint*: The final Gasket dust is left unchanged by this IFS.)

2.3. Consider the following transformations:

> Scale by 0.333.
>
> Scale by 0.333, then translate by +0.667 in the x-direction.
>
> Collage the pieces.

Would these transformations lead to the Cantor Middle Thirds Set if we had started with something besides a nicely oriented line segment (as we did in Figure 2.7)? Try starting with a square, say. Draw a few generations by hand. What do you get?

2.4. Consider the IFS defined by the following rules:

> Scale by 0.5.
>
> Scale by 0.5, then translate by +0.5 in the y-direction.

Scale by 0.5, then rotate (counterclockwise) by $90°$, then translate by $+1$ in the x-direction.

Collage the pieces.

(a) Sketch the result of one application of these rules to the unit square covered with horizontal stripes.

(b) Now sketch the result of one application of these rules to the picture you got in part (a).

2.5. Suppose the first of the three transformations for the Gasket listed on page 44, is rewritten as "scale by 0.5, then rotate around the origin by $120°$ in the counterclockwise direction, then translate by $+0.5$ in the x-direction." The resulting new set of rules gives the same Gasket as before. Can you see why? (*Hint:* what do the new transformations do to an equilateral triangle?) Can you find a set of rules with **four** transformations that produce the same Gasket?

2.6. Referring to Figure 2.6, find the regions in the final dust in which the last transformations were (a) **111**, (b) **123**, (c) **132**, (d) **231**, and (e) **321**, where **1** is "scale by 0.333," **2** is "scale by 0.333 then translate by $+0.5$ in the x-direction," and **3** is "scale by 0.333 then translate by $+0.5$ in the y-direction."

2.7. Let rules **1**, **2**, and **3** respectively be

Scale by 0.5.

Scale by 0.5, then translate by $+0.5$ in the x-direction.

Scale by 0.5, then translate by $+0.5$ in the y-direction.

(a) What picture would you ultimately get from applying the rules in the sequence **1, 2, 1, 2, 1, 2**, . . . ? Start at the point $(0.5, 0.5)$ and plot about 10 of the resulting points to check your answer.

(b) What about the sequence **1, 2, 3, 1, 2, 3, 1, 2, 3**, . . . ?

(c) What about the sequence **1, 1, 2, 3, 1, 1, 2, 3, 1, 1, 2, 3**, . . . ? Note that you can speed up your procedure by substituting the equivalent Chaos Game rules "go $\frac{1}{2}$-way toward $(0, 0)$ [or $(1, 0)$ or (0.1)]."

CHAPTER THREE

3.1. Compute the dimension of the isoceles right Gasket of side length s for any positive number s.

3.2. How disconnected is the Cantor Middle Thirds Set (Cantor MTS)? Here are two simple ways to get some indication of just how badly the Cantor MTS is chopped up.

(a) First, measure the length at each stage in constructing the Cantor MTS. Because the MTS is made by removing intervals, each of our measurements exceeds the length of the MTS, but we can look for a pattern and deduce the actual length. To begin, we cover the MTS with one interval of length 1. Next, we cover it with two intervals of length 1/3, so its length is less than 2/3. Continue this line of reasoning and make some conclusion about the length of the Cantor MTS.

(b) Now measure the lengths of the intervals removed in forming the Cantor MTS and see what remains. For example, first we remove one interval of length 1/3, then two intervals of length 1/9, and so on. Using a calculator, add up the lengths removed $(1/3 + 2/9 + \ldots)$. What do you get for the length of the MTS?

(c) Do your answers for parts (a) and (b) agree?

3.3. (a) Using equation (3) from Chapter Three, compute the dimension of the Cantor set formed from the unit interval by removing the middle half of the interval.

(b) Compute the dimension of the Cantor set formed from the unit interval by removing the middle two-thirds of the interval.

(c) Compute the dimension of the Cantor set formed from the unit interval by removing the middle fifth of the interval.

(d) Compute the dimension of the Cantor set formed from the unit interval by removing the middle segment of length s, where s is any number between 0 and 1.

3.4. Make a series of tightly wadded balls, as described in Chapter Three, of tissue paper (very un-stiff) and of brown-paper-bag paper (very stiff). Before making any measurements, guess which collection has the higher dimension. What reasons can you give to support your answer? Now do the experiment. Were you right or wrong?

3.5. Make a copy of finely divided graph paper on an overhead foil. Use this as one of the sheets in the "blob of goo between two sheets of plastic" experiment described in Chapter Three. Put a small blob of goo on this sheet, cover it with a plain overhead transparency foil, and press firmly on the top foil until the small blob squeezes out into a large, thin blob. (You'll probably want to use water-soluble goo so you can easily wash it off.) Pull the foils apart, producing a fingery pattern. Do box counting to estimate the average dimension of the boundary of pattern. Try the experiment with different pulling speeds. Does the dimension depend on the speed with which the sheets are separated? Also try different "goos." Does the dimension depend on the material used?

3.6. Make a notch in the middle of one side of a piece of typing paper. Hold the sides of the paper firmly and pull them apart (in the plane of the paper) until the paper tears. The tear edge will be jagged and fractal-like. Overlay the torn edge on a piece of finely divided graph paper and estimate the dimension of the edge by box counting. How does the dimension depend on the speed with which the paper is torn? Try tissue paper instead of typing paper. How does the character of the edge depend on the structure of the paper?

CHAPTER FOUR

4.1. You have held an account that has yielded a constant rate of 10% per year for a number of years. Your current balance is $1000. How many years ago did your balance first become as large as $500? Solve the exercise graphically. Can you see how to solve this analytically?

4.2. Two competing banks offer "bonus/service charge" accounts. One offers a bonus of $500 and a service charge rate of 20% (that is, the effective interest rate is −20%), the second a bonus of $300 and a service charge rate of 10%. Which is the better deal in the long run, if you start with a $0 balance?

4.3.   (a) Show that the maximum height of the tent map graph is s/2.

(b) Show that the nonzero fixed point occurs at $x_f = \dfrac{s}{1+s}$.

4.4. Show that the first iterate of s/2 in the tent map has the value $s \cdot (1 - \dfrac{s}{2})$.

4.5. (a) Using the results of Explorations 4.3 and 4.4, above, show that the condition on s for band merging is equivalent to setting $s^3 - s^2 - 2s + 2 = 0$.

(b) Verify that the numerical value of s where the chaotic bands merge in the bifurcation diagram of the tent map is $\sqrt{2}$.

4.6. The popular media contain many references to the Butterfly Effect, often stating or implying that "small causes can produce big consequences." Find such a reference and comment on whether the example that the author uses is valid or reflects a misunderstanding.

4.7. Historians are frequently confronted with the task of trying to discern cause-and-effect relations among seemingly random occurrences. How might the ideas of deterministic chaos alter the study of human history?

4.8. For s > 2 in the tent map, we have seen that the interval [0, 1] contains many points escaping to $-\infty$. Show that the points that do not escape to $-\infty$ form a Cantor Set, and find the similarity dimension of this Cantor Set. *Note*: The dimension will be expressed in terms of the slope s.

CHAPTER FIVE

5.1. Show that the equation of the nonzero fixed point of the logistic map is $x_f = 1 - 1/s$.

5.2. (a) The sequence generated by iterating the logistic map starting from almost all x in the interval [0, 1] converges to the fixed point, $1 - 1/s$, provided s is between 1 and 3. Describe those x for which the sequence does not converge to $1 - 1/s$.

(b) For s between 3 and 3.449 the logistic map has an attracting 2-cycle. Show that there are infinitely many points that do not converge to this 2-cycle.

5.3. Consider the logistic map with $s = 4$. Taking two points near $1/2$, say $x_0 = 0.490$ and $x'_0 = 0.491$, determine what happens to the distance between these points under one iteration of this logistic map. How is this result consistent with sensitive dependence on initial conditions?

5.4. The first few period doublings in the first cascade for the sine map occur at the approximate parameter values 2.2618, 2.6179, 2.6975, 2.7147, and 2.7184. Is the approximate value for the Feigenbaum number you would infer from these data comparable to that determined for the logistic map?

5.5. A dripping faucet can be made to undergo a period-doubling sequence by adjusting the flow rate. See if you can observe at least the first period doubling. At the very lowest drip rate successive drops have a fixed interval. You can hear this periodicity by placing a pie tin under the faucet. Period-one behavior sounds something like PLOP, PLOP, PLOP, ... Very gradually increase the flow rate. Listen for a change in the rhythm of the drops. Period-two behavior sounds like PLOP, plop, PLOP, plop, ... Can you hear the difference between period-2 and chaotic dripping?

CHAPTER SIX

6.1. Suppose we know that the following time series comes from the logistic map?

0.429, 0.960, 0.150, 0.500, 0.980, 0.077, 0.278, 0.787, 0.657, 0.883, 0.405, 0.945, 0.204, 0.637, 0.906, 0.333, 0.871, 0.440, 0.966.

Without doing a delay plot, can you find the parameter value of the logistic map from the data? Do this by a calculation in which you use two successive values of the time series. Can you see a way to find the parameter value using only one (particular) value of the time series?

6.2. Suppose we have a data set, x_1, . . . , x_{20}, made from a 5-cycle x_1, x_2, x_3, x_4, and x_5. (In other words, $x_6 = x_1$, $x_7 = x_2$,

$x_8 = x_3$, and so on.) Suppose we have taken a filter smaller than $|x_1 - x_2|$, $|x_1 - x_3|$, $|x_1 - x_4|$, $|x_1 - x_5|$, $|x_2 - x_3|$, $|x_2 - x_4|$, $|x_2 - x_5|$, $|x_3 - x_4|$, $|x_3 - x_5|$, and $|x_4 - x_5|$. Sketch the close-pairs plot for this data set.

In each of the Explorations 6.3 through 6.5, draw a square IFS picture with corners labeled as in Figures 6.26–6.29.

6.3. Draw the picture if the driving signal is
   (a) the 2-cycle 1, 2, 1, 2, . . . ,
   (b) the 2-cycle 1, 3, 1, 3, . . . ,
   (c) the 3-cycle, 2, 3, 4, 2, 3, 4, . . .

6.4. Suppose the data string is uniformly random, but only on the symbols 1, 2, and 3, that is, 4 never occurs. Sketch the picture you expect from this string.

6.5. Suppose the data string is uniformly random, with the restriction that 2 never follows 1. Sketch the picture you expect from this string.

6.6. Make a 4-color or 4-shade "Kelly painting" from data sorted into bins with values ranging from 0 to 1, from 1 to 2, from 2 to 3, and from 3 to 4. Suppose the signal is the 5-cycle 0.5, 1.5, 0.7, 2.5, 3.5, 0.5, 1.5, 0.7, and so on. Make one picture with an array of 5 rows and 3 columns, a second with 5 rows and 5 columns, and a third with 5 rows and 7 columns. Can you see obvious patterns emerging? How are the patterns related to the relationship between the period of the signal and the number of columns?

6.7. A tent map has a parameter value equal to 1.5. At one instant x is 10% greater than the fixed point value, for this tent map. By how much would you have to change s so that the next value of x would exactly equal that of the fixed point?

6.8. A tent map has a parameter value equal to 1.5. Suppose s can be changed by a maximum of ±10% in an attempt to control x near the nonzero fixed point value. Start with x = 0.7. What sequence of changes in s are required to bring x to within ±5% of the fixed point in the shortest time?

6.9. Make a list of phenomena in your everyday life that are possible examples of chaotic dynamics. For each, speculate on the consequences (good and/or bad) of your being able to control chaos with small perturbations. How about control by others?

7.1. Apply the Mandelbrot map (equations (1a) and (1b) of Chapter Seven) three times to each of these points, always starting with $x_0 = 0$, $y_0 = 0$. (a) $a = 1$, $b = 0$, (b) $a = -1$, $b = 0$, (c) $a = 1$, $b = 1$, (d) $a = -1$, $b = 1$, (e) $a = -1$, $b = 0.25$, (f) $a = -1$, $b = -0.25$.

7.2. For each application of the Mandelbrot map in Exploration 7.1, compute the distance between the point and the origin.

7.3. Using a cut-off time of 3, find which points in Exploration 7.1 lie in the Mandelbrot Set. (*Hint*: The distance calculations were done in Exploration 7.2.)

7.4. For $a < -2$ and $b = 0$ (part iii in Figure 7.18), use graphical iteration to convince yourself that the Julia Set associated with Mandelbrot map intersects the x-axis in only a Cantor Set.

7.5. How does the M Set contribute to our understanding of what is simple and what is complicated? Express this lesson, perhaps metaphorically, through an experience you have had. For example, think of the many social interactions in which you engage that are iterative processes: you do something, the person with whom you are interacting responds, then you respond to that, then . . . .

CHAPTER EIGHT

8.1.   (a) Sketch the graph of the map $x_{n+1} = x_n^3$. (*Hint*: Compute x_{n+1} for $x_n = 0$, $\pm 1/2$, ± 1, $\pm 3/2$, ± 2.)
   (b) Find the fixed points of the map. (There are three.)
   (c) Which of these fixed points are stable? Which are unstable?
   (d) Use graphical iteration to find the basins of attraction of the stable fixed point(s).
   (e) Use graphical iteration to determine the fate of the points that are not in the basins of attraction of the stable fixed point(s).

8.2. Consider the family of functions $f_c(x) = x^2 + c$, for c and x real numbers.

(a) Show that $f_c(x)$ has two real roots when $c < 0$. What are these roots?

(b) Using a calculator, sketch the graph of the Newton map for $c = -1/2$.

(c) Use graphical iteration to determine the fate of initial points under this Newton map.

(d) Repeat parts (b) and (c) for $c = -3/2$.

(e) What do you think happens for any $c < 0$?

(f) Repeat parts (b) and (c) for $c = 1/2$.

(g) What do you think happens for any $c > 0$?

8.3. Suppose for some function $f(x)$ there is a special value of x, x^*, such that $f(x^*) = 0$ and $f'(x^*) \neq 0$. Show that x^* is a fixed point for the associated Newton map.

8.4. (a) Write a parable of your own. Specifically, take an instance from your life where a small choice, perhaps made in an offhand fashion, led eventually to a significant difference in your life. For your parable to be relevant to the context of Chapter Eight, the course of events flowing from your choice has to be deterministic. Try to identify some of the nonlinearities in the "dynamics of your life" that helped cause your choice to initiate the unexpected chain of events it did.

(b) To emphasize the magnitude of the effect of this small choice, construct a plausible scenario of the result of your having made a different small choice (under the same "life dynamics").

8.5. Write a fictional version of Exploration 8.4. You are no longer constrained to stick to the truth. Use your imagination to tell a story that illustrates how small changes can result in arriving at significantly different stable states when the dynamics impelling change is nonlinear.

CHAPTER NINE

Explorations 9.1 and 9.2 use the rule ☐☐■ *or* ■☐☐.

9.1. Determine by hand the first six generations of the starting configuration ☐☐☐☐■■☐■☐☐☐☐ .

9.2. Determine by hand the first six generations of the starting configuration ⬜⬜⬜⬜⬛⬛⬜⬛⬜⬜⬜ .

9.3. Determine by hand the first six generations of an initial single on-cell for the von Neumann automaton with the following rule: "a dead central cell becomes alive in the next generation if exactly 1 of its 4 neighbors (N, S, E, or W) is alive in the current generation."

9.4. **Voting rules** are special cases of outer totalistic rules: for neighborhoods with an odd number of cells (all the examples we have considered satisfy this), the central cell becomes alive if more than half of the current neighborhood cells are alive. Write the outer totalistic formulation of the voting rule for (a) one-dimensional 3-cell neighborhood automata, (b) one-dimensional 5-cell neighborhood automata, (c) two-dimensional von Neumann automata, and (d) two-dimensional Moore automata.

9.5. Get some dried peas. Place a flat dish inside a box (so that the peas don't get everywhere) and pour peas onto it. Note that at first, the pile gets higher and higher but that after a while, the shape of the pile on the dish doesn't change very much. (That's the critical state.) Now, drop one pea at a time near the top of the (critical) pile and observe the resulting avalanches. Do you see that most of the time very little happens but that occasionally a large avalanche breaks loose? Repeat the procedure using dried rice and dried lentils. In each case, observe the shape of the critical pile. Can you draw a conclusion about the angle of repose of a pile and the character of the individual grains from which it is made? Incidentally, rice is especially interesting because, after many avalanches, the rice grains tend to get aligned. In which direction is the alignment? Do you see a similar alignment with lentils?

9.6. Using the neural network with weights shown on the left side of Figure 9.35, test the pair $(p, q)/o = (T, T)/T$. If necessary, use backpropagation to find weights that satisfy this input/output pair. Use the conversion rule given in the example worked out in the text.

9.7. How well do the weights found in Exploration 9.6 compute the remaining entries in the IMPLIES table?

9.8. Here are two opposite views of humanity.

1. The common experiences of self-awareness, free will, and feelings can never be understood in terms of physical laws.

2. People are mechanistic entities whose behaviors emerge from the same principles that govern all forms of matter. Construct an argument that supports each position. Which argument do you find more compelling?

Index